THE ELEMENTS OF EUCLID.

THE FIRST SIX BOOKS

OF THE

ELEMENTS OF EUCLID,

AND

PROPOSITIONS I.-XXI. OF BOOK XI.,

AND AN

APPENDIX ON THE CYLINDER, SPHERE, CONE, ETC.,

WITH

COPIOUS ANNOTATIONS AND NUMEROUS EXERCISES.

BY

JOHN CASEY, LL.D., F.R.S.,

FELLOW OF THE ROYAL UNIVERSITY OF IRELAND;
MEMBER OF COUNCIL, ROYAL IRISH ACADEMY;
MEMBER OF THE MATHEMATICAL SOCIETIES OF LONDON AND FRANCE;
AND PROFESSOR OF THE HIGHER MATHEMATICS AND OF
MATHEMATICAL PHYSICS IN THE CATHOLIC UNIVERSITY OF IRELAND.

THIRD EDITION, REVISED AND ENLARGED.
DUBLIN: HODGES, FIGGIS, & CO., GRAFTON-ST.
LONDON: LONGMANS, GREEN, & CO.
1885.

ISBN: 978-1-963956-18-4

DUBLIN
PRINTED AT THE UNIVERSITY PRESS,
BY PONSONBY AND WELDRICK

PREFACE.

This edition of the Elements of Euclid, undertaken at the request of the principals of some of the leading Colleges and Schools of Ireland, is intended to supply a want much felt by teachers at the present day—the production of a work which, while giving the unrivalled original in all its integrity, would also contain the modern conceptions and developments of the portion of Geometry over which the Elements extend. A cursory examination of the work will show that the Editor has gone much further in this latter direction than any of his predecessors, for it will be found to contain, not only more actual matter than is given in any of theirs with which he is acquainted, but also much of a special character, which is not given, so far as he is aware, in any former work on the subject. The great extension of geometrical methods in recent times has made such a work a necessity for the student, to enable him not only to read with advantage, but even to understand those mathematical writings of modern times which require an accurate knowledge of Elementary Geometry, and to which it is in reality the best introduction.

In compiling his work the Editor has received invaluable assistance from the late Rev. Professor Townsend, S.F.T.C.D. The book was rewritten and considerably altered in accordance with his suggestions, and to that distinguished Geometer it is largely indebted for whatever merit it possesses.

The Questions for Examination in the early part of the First Book are intended as specimens, which the teacher ought to follow through the entire work. Every person who has had experience in tuition knows well the importance of such examinations in teaching Elementary Geometry.

The Exercises, of which there are over eight hundred, have been all selected with great care. Those in the body of each Book are intended as applications of Euclid's Propositions. They are for the most part of an elementary character, and may be regarded as common property, nearly every one of them having appeared already in previous collections. The Exercises at the end of each Book are more advanced; several are due to the late Professor Townsend, some are original, and a large number have been taken from two important French works—CATALAN's *Théorèmes et Problèmes de Géométrie Elémentaire*, and the *Traité de Géométrie*, by ROUCHÉ and DE COMBEROUSSE.

The second edition has been thoroughly revised and greatly enlarged. The new matter includes several alternative proofs, important examination questions on each of the books, an explanation of the ratio of incommensurable quantities, the first twenty-one propositions of Book XI., and an Appendix on the properties of the Prism, Pyramids, Cylinder, Sphere, and Cone.

The present Edition has been very carefully read throughout, and it is hoped that few misprints have escaped detection.

The Editor is glad to find from the rapid sale of former editions (each 3000 copies) of his Book, and its general adoption in schools, that it is likely to

i

accomplish the double object with which it was written, viz. to supply students with a Manual that will impart a thorough knowledge of the immortal work of the great Greek Geometer, and introduce them, at the same time, to some of the most important conceptions and developments of the Geometry of the present day.

JOHN CASEY.

86, SOUTH CIRCULAR-ROAD, DUBLIN.
November, 1885.

Contents

BOOK VI.

BOOK XI.

APPENDIX.

NOTES.

THE ELEMENTS OF EUCLID.

INTRODUCTION.

Geometry is the Science of figured Space. Figured Space is of one, two, or three dimensions, according as it consists of lines, surfaces, or solids. The boundaries of solids are surfaces; of surfaces, lines; and of lines, points. Thus it is the province of Geometry to investigate the properties of solids, of surfaces, and of the figures described on surfaces. The simplest of all surfaces is the plane, and that department of Geometry which is occupied with the lines and curves drawn on a plane is called *Plane Geometry*; that which demonstrates the properties of solids, of curved surfaces, and the figures described on curved surfaces, is *Geometry of Three Dimensions*. The simplest lines that can be drawn on a plane are the right line and circle, and the study of the properties of the point, the right line, and the circle, is the introduction to Geometry, of which it forms an extensive and important department. This is the part of Geometry on which the oldest Mathematical Book in existence, namely, Euclid's *Elements*, is written, and is the subject of the present volume. The *conic sections* and *other curves* that can be described on a plane form special branches, and complete the divisions of this, the most comprehensive of all the Sciences. The student will find in Chasles' *Aperçu Historique* a valuable history of the origin and the development of the methods of Geometry.

In the following work, when figures are not drawn, the student should construct them from the given directions. The Propositions of Euclid will be printed in larger type, and will be referred to by Roman numerals enclosed in brackets. Thus [III. XXXII.] will denote the 32nd Proposition of the 3rd Book. The number of the Book will be given only when different from that under which the reference occurs. The general and the particular enunciation of every Proposition will be given in one. By omitting the letters enclosed in parentheses we have the general enunciation, and by reading them, the particular. The annotations will be printed in smaller type. The following symbols will be used in them:—

Circle	will be denoted by	⊙
Triangle	,,	△
Parallelogram	,,	▱
Parallel lines	,,	‖
Perpendicular	,,	⊥

In addition to these we shall employ the usual symbols $+$, $-$, &c. of Algebra, and also the sign of congruence, namely $\equiv$. This symbol has been introduced by the illustrious Gauss.

BOOK I.

THEORY OF ANGLES, TRIANGLES, PARALLEL LINES, AND PARALLELOGRAMS.

DEFINITIONS.

THE POINT.

I. A point is that which has position but not dimensions.

A geometrical magnitude which has three dimensions, that is, length, breadth, and thickness, is a solid; that which has two dimensions, such as length and breadth, is a surface; and that which has but one dimension is a line. But a point is neither a solid, nor a surface, nor a line; hence it has no dimensions—that is, it has neither length, breadth, nor thickness.

THE LINE.

II. A line is length without breadth.

A line is space of one dimension. If it had any breadth, no matter how small, it would be space of two dimensions; and if in addition it had any thickness it would be space of three dimensions; hence a line has neither breadth nor thickness.

III. The intersections of lines and their extremities are points.

IV. A line which lies evenly between its extreme points is called a straight or right line, such as *AB*.   **A**————————**B**

If a point move without changing its direction it will describe a right line. The direction in which a point moves in called its "*sense*." If the moving point continually changes its direction it will describe a curve; hence it follows that only one right line can be drawn between two points. The following Illustration is due to Professor Henrici:—"If we suspend a weight by a string, the string becomes stretched, and we say it is straight, by which we mean to express that it has assumed a peculiar definite shape. If we mentally abstract from this string all thickness, we obtain the notion of the simplest of all lines, which we call a straight line."

THE PLANE.

V. A surface is that which has length and breadth.

A surface is space of two dimensions. It has no thickness, for if it had any, however small, it would be space of three dimensions.

VI. When a surface is such that the right line joining any two arbitrary points in it lies wholly in the surface, it is called a *plane*.

A plane is perfectly flat and even, like the surface of still water, or of a smooth floor.— NEWCOMB.

FIGURES.

VII. Any combination of points, of lines, or of points and lines in a plane, is called a *plane* figure. If a figure be formed of points only it is called a *stigmatic* figure; and if of right lines only, a *rectilineal* figure.

VIII. Points which lie on the same right line are called *collinear* points. A figure formed of collinear points is called a *row* of points.

THE ANGLE.

IX. The inclination of two right lines extending out from one point in different directions is called a *rectilineal* angle.

X. The two lines are called the *legs*, and the point the *vertex* of the angle.

A right line drawn from the vertex and turning about it in the plane of the angle, from the position of coincidence with one leg to that of coincidence with the other, is said to turn through the angle, and the angle is the greater as the quantity of turning is the greater. Again, since the line may turn from one position to the other in either of two ways, two angles are formed by two lines drawn from a point.

Thus if AB, AC be the legs, a line may turn from the position AB to the position AC in the two ways indicated by the arrows. The smaller of the angles thus formed is to be understood as the angle contained by the lines. The larger, called a *re-entrant* angle, seldom occurs in the "Elements."

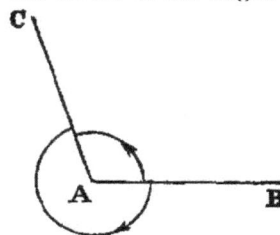

XI. *Designation of Angles.*—A particular angle in a figure is denoted by three letters, as BAC, of which the middle one, A, is at the vertex, and the other two along the legs. The angle is then read BAC.

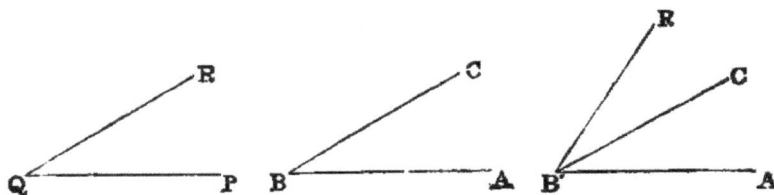

XII. The angle formed by joining two or more angles together is called their *sum.* Thus the sum of the two angles ABC, PQR is the angle $AB'R$, formed by applying the side QP to the side BC, so that the vertex Q shall fall on the vertex B, and the side QR on the opposite side of BC from BA.

XIII. When the sum of two angles BAC, CAD is such that the legs BA, AD form one right line, they are called *supplements* of each other.

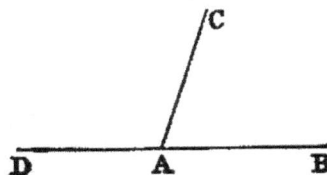

Hence, when one line stands on another, the two angles which it makes on the same side of that on which it stands are supplements of each other.

3

XIV. When one line stands on another, and makes the adjacent angles at both sides of itself equal, each of the angles is called a *right* angle, and the line which stands on the other is called a *perpendicular* to it.

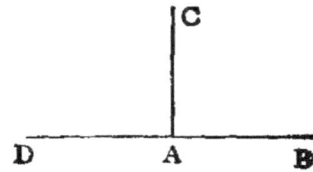

Hence a right angle is equal to its supplement.

XV. An *acute* angle is one which is less than a right angle, as *A*.

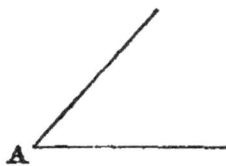

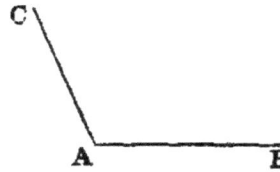

XVI. An *obtuse* angle is one which is greater than a right angle, as *BAC*.

The supplement of an acute angle is obtuse, and conversely, the supplement of an obtuse angle is acute.

XVII. When the sum of two angles is a right angle, each is called the *complement* of the other. Thus, if the angle *BAC* be right, the angles *BAD*, *DAC* are complements of each other.

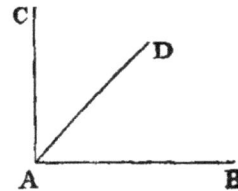

Concurrent Lines.

XVIII. Three or more right lines passing through the same point are called *concurrent* lines.

XIX. A system of more than three concurrent lines is called a *pencil* of lines. Each line of a pencil is called a *ray*, and the common point through which the rays pass is called the *vertex*.

The Triangle.

XX. A *triangle* is a figure formed by three right lines joined end to end. The three lines are called its *sides*.

XXI. A triangle whose three sides are unequal is said to be *scalene*, as *A*; a triangle having two sides equal, to be *isosceles*, as *B*; and and having all its sides equal, to be *equilateral*, as *C*.

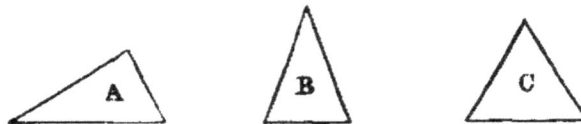

XXII. A *right-angled* triangle is one that has one of its angles a right angle, as *D*. The side which subtends the right angle is called the *hypotenuse*.

XXIII. An *obtuse-angled* triangle is one that has one of its angles obtuse, as *E*.

XXIV. An *acute-angled* triangle is one that has its three angles acute, as *F*.

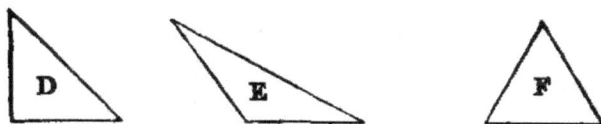

XXV. An *exterior* angle of a triangle is one that is formed by any side and the continuation of another side.

Hence a triangle has six exterior angles; and also each exterior angle is the supplement of the adjacent interior angle.

THE POLYGON.

XXVI. A *rectilineal* figure bounded by more than three right lines is usually called a polygon.

XXVII. A polygon is said to be *convex* when it has no re-entrant angle.

XXVIII. A polygon of four sides is called a *quadrilateral.*

XXIX. A quadrilateral whose four sides are equal is called a *lozenge.*

XXX. A lozenge which has a right angle is called a *square.*

XXXI. A polygon which has five sides is called a *pentagon*; one which has six sides, a *hexagon*, and so on.

THE CIRCLE.

XXXII. A *circle* is a plane figure formed by a curved line called the *circumference*, and is such that all right lines drawn from a certain point within the figure to the circumference are equal to one another. This point is called the *centre.*

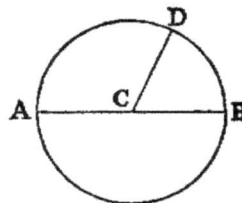

XXXIII. A *radius* of a circle is any right line drawn from the centre to the circumference, such as *CD.*

XXXIV. A *diameter* of a circle is a right line drawn through the centre and terminated both ways by the circumference, such as *AB.*

From the definition of a circle it follows at once that the path of a movable point in a plane which remains at a constant distance from a fixed point is a circle; also that any point *P* in the plane is inside, outside, or on the circumference of a circle according as its distance from the centre is *less* than, *greater* than, or *equal* to, the radius.

POSTULATES.

Let it be granted that—

I. A right line may be drawn from any one point to any other point.

When we consider a straight line contained between two fixed points which are its ends, such a portion is called a *finite straight line.*

II. A terminated right line may be produced to any length in a right line.

5

Every right line may extend without limit in either direction or in both. It is in these cases called an *indefinite* line. By this postulate a finite right line may be supposed to be produced, whenever we please, into an indefinite right line.

III. A circle may be described from any centre, and with any distance from that centre as radius.

If there be two points A and B, and if with any instruments, such as a ruler and pen, we draw a line from A to B, this will evidently have some irregularities, and also some breadth and

A —————————— **B**

thickness. Hence it will not be a geometrical line no matter how nearly it may approach to one. This is the reason that Euclid postulates the drawing of a right line from one point to another. For if it could be accurately done there would be no need for his asking us to let it be granted. Similar observations apply to the other postulates. It is also worthy of remark that Euclid never takes for granted the doing of anything for which a geometrical construction, founded on other problems or on the foregoing postulates, can be given.

AXIOMS.

I. Things which are equal to the same, or to equals, are equal to each other.

Thus, if there be three things, and if the first, and the second, be each equal to the third, we infer by this axiom that the first is equal to the second. This axiom relates to all kinds of magnitude. The same is true of Axioms II., III., IV., V., VI., VII., IX.; but VIII., X., XI., XII., are strictly geometrical.

II. If equals be added to equals the sums will be equal.

III. If equals be taken from equals the remainders will be equal.

IV. If equals be added to unequals the sums will be unequal.

V. If equals be taken from unequals the remainders will be unequal.

VI. The doubles of equal magnitudes are equal.

VII. The halves of equal magnitudes are equal.

VIII. Magnitudes that can be made to coincide are equal.

The placing of one geometrical magnitude on another, such as a line on a line, a triangle on a triangle, or a circle on a circle, &c., is called *superposition*. The superposition employed in Geometry is only *mental*, that is, we conceive one magnitude placed on the other; and then, if we can prove that they coincide, we infer, by the present axiom, that they are equal. Superposition involves the following principle, of which, without explicitly stating it, Euclid makes frequent use:—"Any figure may be transferred from one position to another without change of form or size."

IX. The whole is greater than its part.

This axiom is included in the following, which is a fuller statement:—

IX′. The whole is equal to the sum of all its parts.

X. Two right lines cannot enclose a space.

This is equivalent to the statement, "If two right lines have two points common to both, they coincide in direction," that is, they form but one line, and this holds true even when one of the points is at infinity.

XI. All right angles are equal to one another.

This can be proved as follows:—Let there be two right lines AB, CD, and two perpendiculars to them, namely, EF, GH, then if AB, CD be made to coincide by superposition, so that the point E will coincide with G; then since a right angle is equal to its supplement, the line EF must coincide with GH. Hence the angle AEF is equal to CGH.

XII. If two right lines (AB, CD) meet a third line (AC), so as to make the sum of the two interior angles (BAC, ACD) on the same side less than two right angles, these lines being produced shall meet at some *finite distance*.

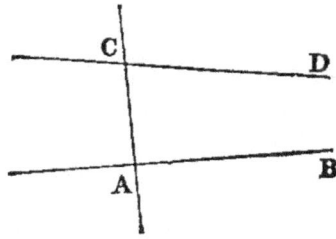

This axiom is the converse of Prop. XVII., Book I.

Explanation of Terms.

Axioms.—"Elements of human reason," according to DUGALD STEWART, are certain general propositions, the truths of which are self-evident, and which are so fundamental, that they cannot be inferred from any propositions which are more elementary; in other words, they are incapable of demonstration. "That two sides of a triangle are greater than the third" is, perhaps, self-evident; but it is not an axiom, inasmuch as it can be inferred by demonstration from other propositions; but we can give no proof of the proposition that "things which are equal to the same are equal to one another," and, being self-evident, it is an axiom.

Propositions which are not axioms are properties of figures obtained by processes of reasoning. They are divided into theorems and problems.

A *Theorem* is the formal statement of a property that may be demonstrated from known propositions. These propositions may themselves be theorems or axioms. A theorem consists of two parts, the *hypothesis*, or that which is assumed, and the *conclusion*, or that which is asserted to follow therefrom. Thus, in the typical theorem,

$$\text{If } X \text{ is } Y, \text{ then } Z \text{ is } W, \qquad \text{(I.)}$$

the hypothesis is that X is Y, and the conclusion is that Z is W.

Converse Theorems.—Two theorems are said to be converse, each of the other, when the hypothesis of either is the conclusion of the other. Thus the converse of the theorem (I.) is—

$$\text{If } Z \text{ is } W, \text{ then } X \text{ is } Y. \qquad \text{(II.)}$$

From the two theorems (I.) and (II.) we may infer two others, called their *contrapositives*. Thus the contrapositive

$$\text{of (I.) is, If } Z \text{ is not } W, \text{ then } X \text{ is not } Y; \qquad \text{(III.)}$$

$$\text{of (II.) is, If } X \text{ is not } Y, \text{ then } Z \text{ is not } W. \qquad \text{(IV.)}$$

The theorem (IV.) is called the *obverse* of (I.), and (III.) the obverse of (II.).

A *Problem* is a proposition in which something is proposed to be done, such as a line to be drawn, or a figure to be constructed, under some given conditions.

7

The *Solution* of a problem is the method of construction which accomplishes the required end.

The *Demonstration* is the proof, in the case of a theorem, that the conclusion follows from the hypothesis; and in the case of a problem, that the construction accomplishes the object proposed.

The *Enunciation* of a problem consists of two parts, namely, the *data*, or things supposed to be given, and the *quaesita*, or things required to be done.

Postulates are the elements of geometrical construction, and occupy the same relation with respect to problems as axioms do to theorems.

A *Corollary* is an inference or deduction from a proposition.

A *Lemma* is an auxiliary proposition required in the demonstration of a principal proposition.

A *Secant* or *Transversal* is a line which cuts a system of lines, a circle, or any other geometrical figure.

Congruent figures are those that can be made to coincide by superposition. They agree in shape and size, but differ in position. Hence it follows, by Axiom VIII., that corresponding parts or portions of congruent figures are congruent, and that congruent figures are equal in every respect.

Rule of Identity.—Under this name the following principle will be sometimes referred to:—"If there is but one X and one Y, then, from the fact that X is Y, it necessarily follows that Y is X."—SYLLABUS.

PROP. I.—PROBLEM.

On a given finite right line (AB) to construct an equilateral triangle.

Sol.—With A as centre, and AB as radius, describe the circle BCD (Post. III.). With B as centre, and BA as radius, describe the circle ACE, cutting the former circle in C. Join CA, CB (Post. I.). Then ABC *is the equilateral triangle required.*

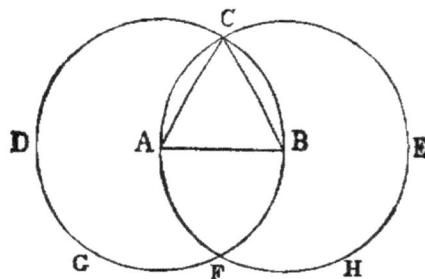

Dem.—Because A is the centre of the circle BCD, AC is equal to AB (Def. XXXII.). Again, because B is the centre of the circle ACE, BC is equal to BA. Hence we have proved.

$$AC = AB,$$
and $$BC = AB.$$

But things which are equal to the same are equal to one another (Axiom I.); therefore AC is equal to BC; therefore the three lines AB, BC, CA are equal to one another. Hence the triangle ABC is equilateral (Def. XXI.); and it is described on the given line AB, *which was required to be done.*

8

Questions for Examination.

1. What is the *datum* in this proposition?
2. What is the *quaesitum*?
3. What is a finite right line?
4. What is the opposite of finite?
5. In what part of the construction is the third postulate quoted? and for what purpose? Where is the first postulate quoted?
6. Where is the first axiom quoted?
7. What use is made of the definition of a circle? What is a circle?
8. What is an equilateral triangle?

Exercises.

The following exercises are to be solved when the pupil has mastered the First Book:—
1. If the lines AF, BF be joined, the figure $ACBF$ is a lozenge.
2. If AB be produced to D and E, the triangles CDF and CEF are equilateral.
3. If CA, CB be produced to meet the circles again in G and H, the points G, F, H are collinear, and the triangle GCH is equilateral.
4. If CF be joined, $CF^2 = 3AB^2$.
5. Describe a circle in the space ACB, bounded by the line AB and the two circles.

PROP. II.—Problem.

From a given point (A) to draw a right line equal to a given finite right line (BC).

Sol.—Join AB (Post. I.); on AB describe the equilateral triangle ABD [I.]. With B as centre, and BC as radius, describe the circle ECH (Post III.). Produce DB to meet the circle ECH in E (Post. II.). With D as centre, and DE as radius, describe the circle EFG (Post. III.). Produce DA to meet this circle in F. AF *is equal to BC.*

Dem.—Because D is the centre of the circle EFG, DF is equal to DE (Def. XXXII.). And because DAB is an equilateral triangle, DA is equal to DB (Def. XXI.). Hence we have

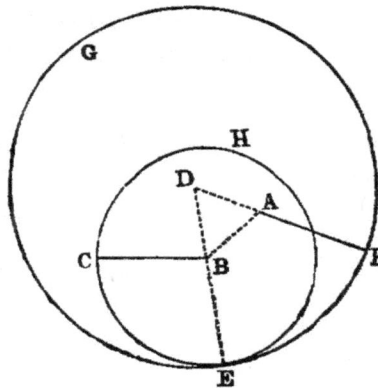

$$DF = DE,$$
and
$$DA = DB;$$

and taking the latter from the former, the remainder AF is equal to the remainder BE (Axiom III.). Again, because B is the centre of the circle ECH, BC is equal to BE; and we have proved that AF is equal to BE; and things which are equal to the same thing are equal to one another (Axiom I.). Hence AF

9

is equal to *BC*. *Therefore from the given point A the line AF has been drawn equal to BC.*

It is usual with commentators on Euclid to say that he allows the use of the *rule* and *compass*. Were such the case this Proposition would have been unnecessary. The fact is, Euclid's object was to teach Theoretical and not Practical Geometry, and the only things he postulates are the drawing of right lines and the describing of circles. If he allowed the mechanical use of the rule and compass he could give methods of solving many problems that go beyond the limits of the "geometry of the point, line, and circle."—*See* Notes D, F at the end of this work.

Exercises.

1. Solve the problem when the point *A* is in the line *BC* itself.
2. Inflect from a given point *A* to a given line *BC* a line equal to a given line. State the number of solutions.

PROP. III.—PROBLEM.

From the greater (AB) of two given right lines to cut off a part equal to (C) the less.

Sol.—From *A*, one of the extremities of *AB*, draw the right line *AD* equal to *C* [II.]; and with *A* as centre, and *AD* as radius, describe the circle *EDF* (Post. III.) cutting *AB* in *E*. *AE* shall be equal to *C*.

Dem.—Because *A* is the centre of the circle *EDF*, *AE* is equal to *AD* (Def. XXXII.), and *C* is equal to *AD* (const.); and things which are equal to the same are equal to one another (Axiom I.); therefore *AE* is equal to *C*. Wherefore *from AB, the greater of the two given lines, a part, AE, has been cut off equal to C, the less.*

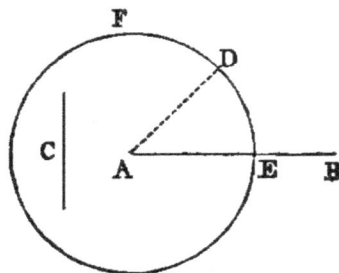

Questions for Examination.

1. What previous problem is employed in the solution of this?
2. What postulate?
3. What axiom in the demonstration?
4. Show how to produce the less of two given lines until the whole produced line becomes equal to the greater.

PROP. IV.—THEOREM.

If two triangles (BAC, EDF) have two sides (BA, AC) of one equal respectively to two sides (ED, DF) of the other, and have also the angles (A, D) included by those sides equal, the triangles shall be equal in every respect—that is, their bases or third sides (BC, EF) shall be equal, and the angles (B, C) at the base of one shall be respectively equal to the angles (E, F) at the base of the other; namely, those shall be equal to which the equal sides are opposite.

10

Dem.—Let us conceive the triangle BAC to be applied to EDF, so that the point A shall coincide with D, and the line AB with DE, and that the point C shall be on the same side of DE as F; then because AB is equal to DE, the point B shall coincide with E. Again, because the angle BAC is equal to the angle EDF, the line AC shall coincide with DF; and since AC is equal to DF (hyp.), the point C shall coincide with F; and we have proved that the point B coincides with E. Hence two points of the line BC coincide with two points of the line EF; and since two right lines cannot enclose a space, BC must coincide with EF. Hence the triangles agree in every respect; *therefore BC is equal to EF, the angle B is equal to the angle E, the angle C to the angle F, and the triangle BAC to the triangle EDF.*

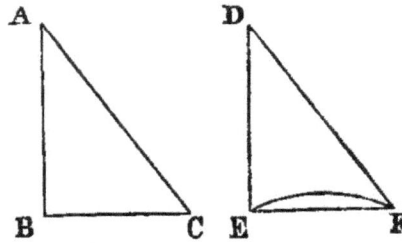

Questions for Examination.

1. How many parts in the hypothesis of this Proposition? *Ans.* Three. Name them.
2. How many in the conclusion? Name them.
3. What technical term is applied to figures which agree in everything but position? *Ans.* They are said to be congruent.
4. What is meant by superposition?
5. What axiom is made use of in superposition?
6. How many parts in a triangle? *Ans.* Six; namely, three sides and three angles.
7. When it is required to prove that two triangles are congruent, how many parts of one must be given equal to corresponding parts of the other? *Ans.* In general, any three except the three angles. This will be established in Props. VIII. and XXVI., taken along with IV.
8. What property of two lines having two common points is quoted in this Proposition? They must coincide.

Exercises.

1. The line that bisects the vertical angle of an isosceles triangle bisects the base perpendicularly.
2. If two adjacent sides of a quadrilateral be equal, and the diagonal bisects the angle between them, their other sides are equal.
3. If two lines be at right angles, and if each bisect the other, then any point in either is equally distant from the extremities of the other.
4. If equilateral triangles be described on the sides of any triangle, the distances between the vertices of the original triangle and the opposite vertices of the equilateral triangles are equal. (This Proposition should be proved after the student has read Prop. XXXII.)

PROP. V.—THEOREM.

The angles (ABC, ACB) at the base (BC) of an isosceles triangle are equal to one another, and if the equal sides (AB, AC) be produced, the external angles (DEC, ECB) below the base shall be equal.

11

Dem.—In BD take any point F, and from AE, the greater, cut off AG equal to AF [III.]. Join BG, CF (Post. I.). Because AF is equal to AG (const.), and AC is equal to AB (hyp.), the two triangles FAC, GAB have the sides FA, AC in one respectively equal to the sides GA, AB in the other; and the included angle A is common to both triangles. Hence [IV.] the base FC is equal to GB, the angle AFC is equal to AGB, and the angle ACF is equal to the angle ABG.

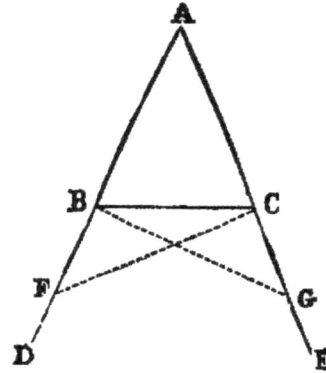

Again, because AF is equal to AG (const.), and AB to AC (hyp.), the remainder, BF, is equal to CG (Axiom III); and we have proved that FC is equal to GB, and the angle BFC equal to the angle CGB. Hence the two triangles BFC, CGB have the two sides BF, FC in one equal to the two sides CG, GB in the other; and the angle BFC contained by the two sides of one equal to the angle CGB contained by the two sides of the other. Therefore [IV.] these triangles have the angle FBC equal to the angle GCB, *and these are the angles below the base*. Also the angle FCB equal to GBC; but the whole angle FCA has been proved equal to the whole angle GBA. Hence the remaining angle ACB is equal to the remaining angle ABC, *and these are the angles at the base*.

Observation.—The great difficulty which beginners find in this Proposition is due to the fact that the two triangles ACF, ABG overlap each other. The teacher should make these triangles separate, as in the annexed diagram, and point out the corresponding parts thus:—

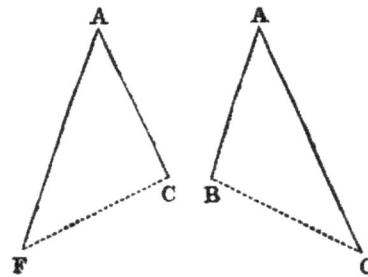

$$AF = AG,$$
$$AC = AB;$$
$$\text{angle } FAC = \text{angle } GAB.$$

Hence [IV.],   angle ACF = angle ABG.
and   angle AFC = angle AGB.

The student should also be shown how to apply one of the triangles to the other, so as to bring them into coincidence. Similar Illustrations may be given of the triangles BFC, CGB.

The following is a very easy proof of this Proposition. Conceive the $\triangle\ ACB$ to be turned, without alteration, round the line AC, until it falls on the other side. Let ACD be its new position; then the angle ADC of the displaced triangle is evidently equal to the angle ABC, with which it originally coincided. Again, the two $\triangle s\ BAC$, CAD have the sides BA, AC of one respectively equal to the sides AC, AD of the other, and the included angles equal; therefore [IV.] the angle ACB opposite to the side AB is equal to the angle ADC opposite to the side AC; but the angle ADC is equal to ABC; therefore ACB is equal to ABC.

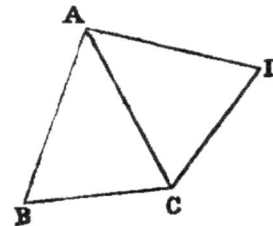

Cor.—Every equilateral triangle is equiangular.

DEF.—*A line in any figure, such as AC in the preceding diagram, which is such that, by folding the plane of the figure round it, one part of the diagram will coincide with the other, is called an* AXIS OF SYMMETRY *of the figure.*

Exercises.

1. Prove that the angles at the base are equal without producing the sides. Also by producing the sides through the vertex.

2. Prove that the line joining the point A to the intersection of the lines CF and BG is an axis of symmetry of the figure.

3. If two isosceles triangles be on the same base, and be either at the same or at opposite sides of it, the line joining their vertices is an axis of symmetry of the figure formed by them.

4. Show how to prove this Proposition by assuming as an axiom that every angle has a bisector.

5. Each diagonal of a lozenge is an axis of symmetry of the lozenge.

6. If three points be taken on the sides of an equilateral triangle, namely, one on each side, at equal distances from the angles, the lines joining them form a new equilateral triangle.

PROP. VI.—THEOREM.

If two angles (B, C) of a triangle be equal, the sides (AC, AB) opposite to them are also equal.

Dem.—If AB, AC are not equal, one must be greater than the other. Suppose AB is the greater, and that the part BD is equal to AC. Join CD (Post. I.). Then the two triangles DBC, ACB have BD equal to AC, and BC common to both. Therefore the two sides DB, BC in one are equal to the two sides AC, CB in the other; and the angle DBC in one is equal to the angle ACB in the other (hyp). Therefore [IV.] the triangle DBC is equal to the triangle ACB—the less to the greater, which is absurd; *hence AC, AB are not unequal, that is, they are equal.*

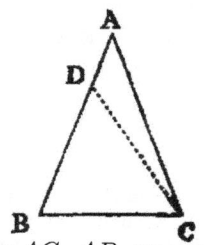

Questions for Examination.

1. What is the hypothesis in this Proposition?
2. What Proposition is this the converse of?
3. What is the obverse of this Proposition?
4. What is the obverse of Prop. v.?
5. What is meant by an indirect proof?
6. How does Euclid generally prove converse Propositions?
7. What false assumption is made in the demonstration?
8. What does this assumption lead to?

PROP. VII—THEOREM.

If two triangles (ACB, ADB) on the same base (AB) and on the same side of it have one pair of conterminous sides (AC, AD) equal to one another, the other pair of conterminous sides (BC, BD) must be unequal.

Dem.—1. Let the vertex of each triangle be without the other. Join CD. Then because AD is equal to AC (hyp.), the triangle ACD is isosceles; therefore [v.] the angle ACD is equal to the angle ADC; but ADC is greater than BDC (Axiom IX.); therefore ACD is greater than BDC: much, more is BCD greater than BDC. Now if the side BD were equal to BC, the angle BCD would be equal to BDC [v.]; but it has been proved to be greater. *Hence BD is not equal to BC.*

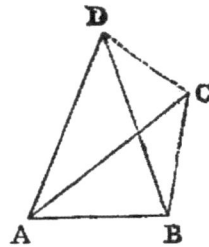

2. Let the vertex of one triangle ADB fall within the other triangle ACB. Produce the sides AC, AD to E and F. Then because AC is equal to AD (hyp.), the triangle ACD is isosceles, and [v.] the external angles ECD, FDC at the other side of the base CD are equal; but ECD is greater than BCD (Axiom IX.). Therefore FDC is greater than BCD: much more is BDC greater than BCD; but if BC were equal to BD, the angle BDC would be equal to BCD [v.]; *therefore BC cannot be equal to BD.*

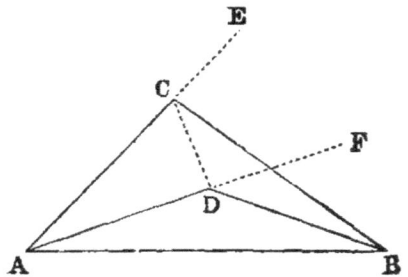

3. If the vertex D of the second triangle fall on the line BC, it is evident that BC and BD are unequal.

Questions for Examination.

1. What use is made of Prop. VII.? *Ans.* As a lemma to Prop. VIII.

2. In the demonstration of Prop. VII. the contrapositive of Prop. V. occurs; show where.

3. Show that two circles can intersect each other only in one point on the same side of the line joining their centres, and hence that two circles cannot have more than two points of intersection.

PROP. VIII.—THEOREM.

If two triangles (ABC, DEF) have two sides (AB, AC) of one respectively equal to two sides (DE, DF) of the other, and have also the base (BC) of one equal to the base (EF) of the other; then the two triangles shall be equal, and the angles of one shall be respectively equal to the angles of the other—namely, those shall be equal to which the equal sides are opposite.

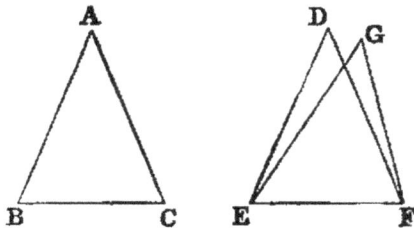

Dem.—Let the triangle ABC be applied to DEF, so that the point B will coincide with E, and the line BC with the line EF; then because BC is equal to EF, the point C shall coincide with F. Then if the vertex A fall on the same side of EF as the vertex D, the point A must coincide with D; for if not, let it take a different position G; then we have EG equal to BA, and BA is equal to ED (hyp.). Hence (Axiom I.) EG is equal to ED: in like manner, FG is equal to FD, and this is impossible [VII.]. Hence the point A must coincide with D, and the triangle ABC agrees in every respect with the triangle DEF; *and therefore the three angles of one are respectively equal to the three angles of the other—namely, A to D, B to E, and C to F, and the two triangles are equal.*

This Proposition is the converse of IV., and is the second case of the congruence of triangles in the Elements.

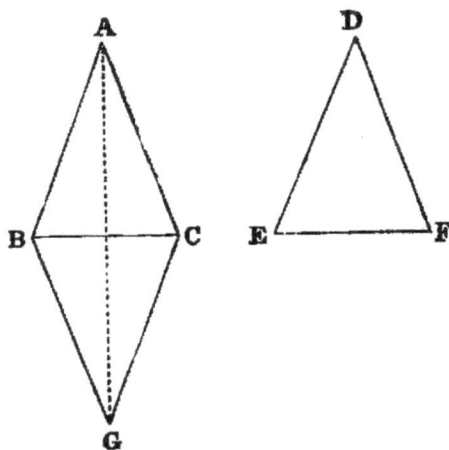

Philo's Proof.—Let the equal bases be applied as in the foregoing proof, but let the vertices be on the opposite sides; then let BGC be the position which EDF takes. Join AG. Then because $BG = BA$, the angle $BAG = BGA$. In like manner the angle $CAG = CGA$. Hence the whole angle $BAC = BGC$; but $BGC = EDF$ therefore $BAC = EDF$.

PROP. IX.—PROBLEM.

To bisect a given rectilineal angle (BAC).

Sol.—In AB take any point D, and cut off [III.] AE equal to AD. Join DE (Post. I.), and upon it, on the side remote from A, describe the equilateral triangle DEF [I.] Join AF. *AF bisects the given angle BAC.*

Dem.—The triangles DAF, EAF have the side AD equal to AE (const.) and AF common; therefore the two sides DA, AF are respectively equal to EA, AF, and the base DF is equal to the base EF, because they are the sides of an equilateral triangle (Def. XXI.). Therefore [VIII.] the angle DAF is equal to the angle EAF; *hence the angle BAC is bisected by the line AF.*

Cor.—The line AF is an axis of symmetry of the figure.

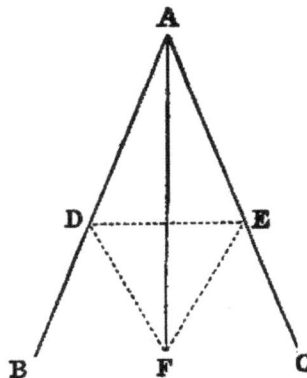

15

Questions for Examination.

1. Why does Euclid describe the equilateral triangle on the side remote from A?
2. In what case would the construction fail, if the equilateral triangle were described on the other side of DE?

Exercises.

1. Prove this Proposition without using Prop. VIII.
2. Prove that AF is perpendicular to DE.
3. Prove that any point in AF is equally distant from the points D and E.
4. Prove that any point in AF is equally distant from the lines AB, AC.

PROP. X.—PROBLEM.

To bisect a given finite right line (AB).

Sol.—Upon AB describe an equilateral triangle ACB [I.]. Bisect the angle ACB by the line CD [IX.], meeting AB in D, *then AB is bisected in D.*

Dem.—The two triangles ACD, BCD, have the side AC equal to BC, being the sides of an equilateral triangle, and CD common. Therefore the two sides AC, CD in one are equal to the two sides BC, CD in the other; and the angle ACD is equal to the angle BCD (const.). Therefore the base AD is equal to the base DB [IV.]. *Hence AB is bisected in D.*

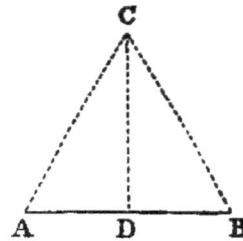

Exercises.

1. Show how to bisect a finite right line by describing two circles.
2. Every point equally distant from the points A, B is in the line CD.

PROP. XI.—PROBLEM.

From a given point (C) *in a given right line* (AB) *to draw a right line perpendicular to the given line.*

Sol.—In AC take any point D, and make CE equal to CD [III.]. Upon DE describe an equilateral triangle DFE [I.]. Join CF. *Then CF shall be at right angles to AB.*

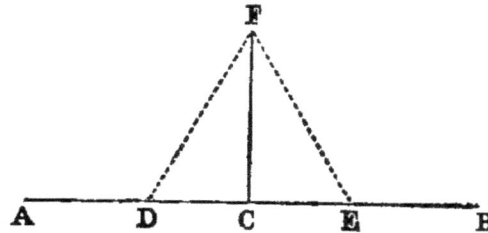

Dem.—The two triangles DCF, ECF have CD equal to CE (const.) and CF common; therefore the two sides CD, CF in one are respectively equal to the two sides CE, CF in the other, and the base DF is equal to the base EF, being the sides of an equilateral triangle (Def. XXI.); therefore [VIII.] the angle DCF is equal to the angle ECF, and they are adjacent angles. Therefore (Def. XIII.) each of them is a right angle, *and CF is perpendicular to AB at the point C.*

16

Exercises.

1. The diagonals of a lozenge bisect each other perpendicularly.
2. Prove Prop. XI. without using Prop. VIII.
3. Erect a line at right angles to a given line at one of its extremities without producing the line.
4. Find a point in a given line that shall be equally distant from two given points.
5. Find a point in a given line such that, if it be joined to two given points on opposite sides of the line, the angle formed by the joining lines shall be bisected by the given line.
6. Find a point that shall be equidistant from three given points.

PROP. XII.—PROBLEM.

To draw a perpendicular to a given indefinite right line (AB) from a given point (C) without it.

Sol.—Take any point D on the other side of AB, and describe (Post. III.) a circle, with C as centre, and CD as radius, meeting AB in the points F and G. Bisect FG in H [x.]. Join CH (Post. I.). CH *shall be at right angles to AB.*

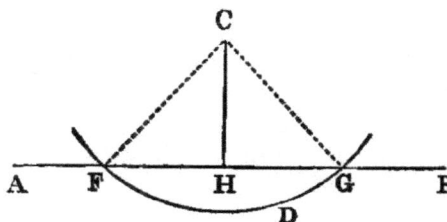

Dem.—Join CF, CG. Then the two triangles FHC, GHC have FH equal to GH (const.), and HC common; and the base CF equal to the base CG, being radii of the circle FDG (Def. XXXII.). Therefore the angle CHF is equal to the angle CHG [VIII.], and, being adjacent angles, they are right angles (Def. XIII.). *Therefore CH is perpendicular to AB.*

Exercises.

1. Prove that the circle cannot meet AB in more than two points.
2. If one angle of a triangle be equal to the sum of the other two, the triangle can be divided into the sum of two isosceles triangles, and the base is equal to twice the line from its middle point to the opposite angle.

PROP. XIII.—THEOREM.

The adjacent angles (ABC, ABD) which one right line (AB) standing on another (CD) makes with it are either both right angles, or their sum is equal to two right angles.

17

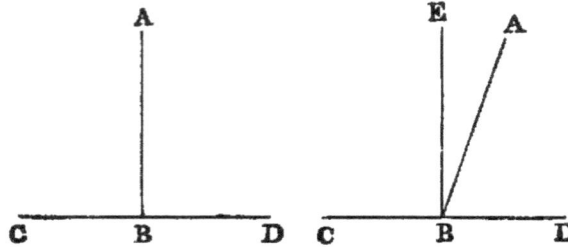

Fig. 1.   Fig. 2.

Dem.—If *AB* is perpendicular to *CD*, as in fig. 1, the angles *ABC*, *ABD* are right angles. If not, draw *BE* perpendicular to *CD* [XI.]. Now the angle *CBA* is equal to the sum of the two angles *CBE*, *EBA* (Def. XI.). Hence, adding the angle *ABD*, the sum of the angles *CBA*, *ABD* is equal to the sum of the three angles *CBE*, *EBA*, *ABD*. In like manner, the sum of the angles *CBE*, *EBD* is equal to the sum of the three angles *CBE*, *EBA*, *ABD*. And things which are equal to the same are equal to one another. Therefore the sum of the angles *CBA*, *ABD* is equal to the sum of the angles *CBE*, *EBD*; but *CBE*, *EBD* are right angles; *therefore the sum of the angles CBA, ABD is two right angles.*

Or thus: Denote the angle *EBA* by θ; then evidently

the angle $\qquad CBA = \text{right angle} + \theta$;

the angle $\qquad ABD = \text{right angle} - \theta$;

therefore $\qquad CBA + ABD = \text{two right angles}.$

Cor. 1.—The sum of two supplemental angles is two right angles.

Cor. 2.—Two right lines cannot have a common segment.

Cor. 3.—The bisector of any angle bisects the corresponding re-entrant angle.

Cor. 4.—The bisectors of two supplemental angles are at right angles to each other.

Cor. 5.—The angle *EBA* is half the difference of the angles *CBA*, *ABD*.

PROP. XIV.–THEOREM.

If at a point (B) in a right line (BA) two other right lines (CB, BD) on opposite sides make the adjacent angles (CBA, ABD) together equal to two right angles, these two right lines form one continuous line.

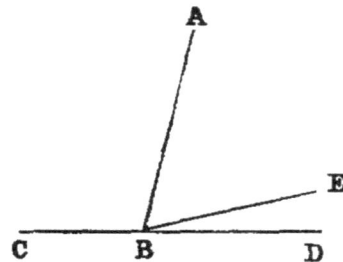

Dem.—If *BD* be not the continuation of *CB*, let *BE* be its continuation. Now, since *CBE* is a right line, and *BA* stands on it, the sum of the angles *CBA*, *ABE* is two right angles (XIII.); and the sum of the angles *CBA*, *ABD* is two right angles (hyp.);

18

therefore the sum of the angles CBA, ABE is equal to the sum of the angles CBA, ABD. Reject the angle CBA, which is common, and we have the angle ABE equal to the angle ABD—that is, a part equal to the whole—which is absurd. *Hence BD must be in the same right line with CB.*

PROP. XV.—THEOREM.

If two right lines (AB, CD) intersect one another, the opposite angles are equal (CEA = DEB, and BEC = AED).

Dem.—Because the line AE stands on CD, the sum of the angles CEA, AED is two right angles [XIII.]; and because the line CE stands on AB, the sum of the angles BEC, CEA is two right angles; therefore the sum of the angles CEA, AED is equal to the sum of the angles BEC, CEA. Reject the angle CEA, which is common, and we have *the angle AED equal to BEC*. In like manner, *the angle CEA is equal to DEB*.

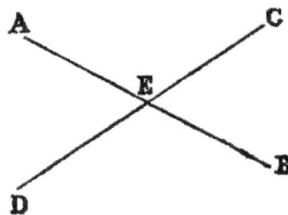

The foregoing proof may be briefly given, by saying that opposite angles are equal because they have a common supplement.

Questions for Examination on Props. XIII., XIV., XV.

1. What problem is required in Euclid's proof of Prop. XIII.?
2. What theorem? *Ans.* No theorem, only the axioms.
3. If two lines intersect, how many pairs of supplemental angles do they make?
4. What relation does Prop. XIV. bear to Prop. XIII.?
5. What three lines in Prop. XIV. are concurrent?
6. What caution is required in the enunciation of Prop. XIV.?
7. State the converse of Prop. XV. Prove it.
8. What is the subject of Props. XIII., XIV., XV.? *Ans.* Angles at a point.

PROP. XVI.—THEOREM.

If any side (BC) of a triangle (ABC) be produced, the exterior angle (ACD) is greater than either of the interior non-adjacent angles.

Dem.—Bisect AC in E [X.]. Join BE (Post. I.). Produce it, and from the produced part cut off EF equal to BE [III]. Join CF. Now because EC is equal to EA (const.), and EF is equal to EB, the triangles CEF, AEB have the sides CE, EF in one equal to the sides AE, EB in the other; and the angle CEF equal to AEB [XV.]. Therefore [IV.] the angle ECF is equal to EAB; but the angle ACD is greater than ECF; therefore the angle ACD is greater than EAB.

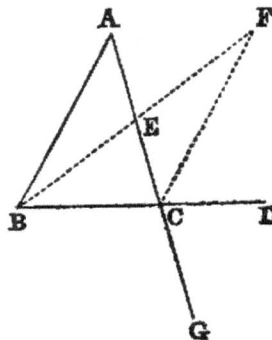

19

In like manner it may be shown, if the side AC be produced, that the exterior angle BCG is greater than the angle ABC; but BCG is equal to ACD [xv.]. Hence ACD is greater than ABC. *Therefore ACD is greater than either of the interior non-adjacent angles A or B of the triangle ABC.*

Cor. 1.—The sum of the three interior angles of the triangle BCF is equal to the sum of the three interior angles of the triangle ABC.

Cor. 2.—The area of BCF is equal to the area of ABC.

Cor. 3.—The lines BA and CF, if produced, cannot meet at any finite distance. For, if they met at any finite point X, the triangle CAX would have an exterior angle BAC equal to the interior angle ACX.

PROP. XVII.—Theorem.

*Any two angles (B, C) of a triangle
(ABC) are together less than two right angles.*

Dem.—Produce BC to D; then the exterior angle ACD is greater than ABC [xvi.]: to each add the angle ACB, and we have the sum of the angles ACD, ACB greater than the sum of the angles ABC, ACB; but the sum of the angles ACD, ACB is two right angles [xiii.]. *Therefore the sum of the angles ABC, ACB is less than two right angles.*

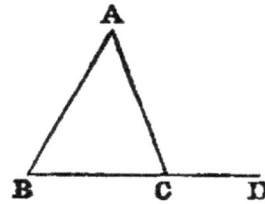

In like manner we may show that the sum of the angles A, B, or of the angles A, C, is less than two right angles.

Cor. 1.—Every triangle must have at least two acute angles.

Cor. 2.—If two angles of a triangle be unequal, the lesser must be acute.

Exercise.

Prove Prop. xvii. without producing a side.

PROP. XVIII.—Theorem.

If in any triangle (ABC) one side (AC) be greater than another (AB), the angle opposite to the greater side is greater than the angle opposite to the less.

Dem.—From AC cut off AD equal to AB [III]. Join BD (Post. I.). Now since AB is equal to AD, the triangle ABD is isosceles; therefore [V.] the angle ADB is equal to ABD; but the angle ADB is greater than the angle ACB [XVI.]; therefore ABD is greater than ACB. *Much more is the angle ABC greater than the angle ACB.*

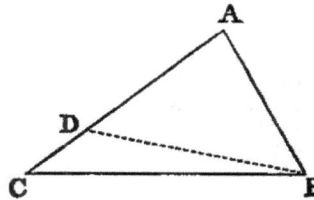

Or thus: From A as centre, with the lesser side AB as radius, describe the circle BED, cutting BC in E. Join AE. Now since AB is equal to AE, the angle AEB is equal to ABE; but AEB is greater than ACB (XVI.); *therefore ABE is greater than ACB.*

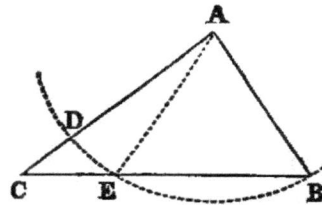

Exercises.

1. If in the second method the circle cut the line CB produced through B, prove the Proposition.

2. This Proposition may be proved by producing the less side.

3. If two of the opposite sides of a quadrilateral be respectively the greatest and least, the angles adjacent to the least are greater than their opposite angles.

4. In any triangle, the perpendicular from the vertex opposite the side which is not less than either of the remaining sides falls within the triangle.

PROP. XIX.—THEOREM.

If one angle (B) of a triangle (ABC) be greater than another angle (C), the side (AC) which it opposite to the greater angle is greater than the side (AB) which is opposite to the less.

Dem.—If AC be not greater than AB, it must be either equal to it or less than it. Let us examine each case:—

1. If AC were equal to AB, the triangle ACB would be isosceles, and then the angle B would be equal to C [V.]; but it is not by hypothesis; therefore AB is not equal to AC.

2. If AC were less than AB, the angle B would be less than the angle C [XVIII.]; but it is not by hypothesis; therefore AC is not less than AB; and since AC is neither equal to AB nor less than it, *it must be greater.*

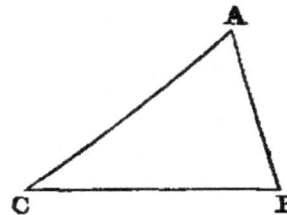

Exercises.

1. Prove this Proposition by a direct demonstration.

2. A line from the vertex of an isosceles triangle to any point in the base is less than either of the equal sides, but greater if the point be in the base produced.

3. Three equal lines could not be drawn from the same point to the same line.

21

4. The perpendicular is the least line which can be drawn from a given point to a given line; and of all others that may be drawn to it, that which is nearest to the perpendicular is less than any one more remote.

5. If in the fig., Prop. XVI., AB be the greatest side of the $\triangle$ ABC, BF is the greatest side of the $\triangle$ FBC, and the angle BFC is less than half the angle ABC.

6. If ABC be a $\triangle$ having AB not greater than AC, a line AG, drawn from A to any point G in BC, is less than AC. For the angle ACB [XVIII.] is not greater than ABC; but AGC [XVI.] is greater than ABC; therefore AGC is greater than ACG. *Hence AC is greater than AG.*

PROP. XX.—Theorem.

The sum of any two sides (BA, AC) of a triangle (ABC) is greater than the third.

Dem.—Produce BA to D (Post. II.), and make AD equal to AC [III.]. Join CD. Then because AD is equal to AC, the angle ACD is equal to ADC (V.); therefore the angle BCD is greater than the angle BDC; hence the side BD opposite to the greater angle is greater than BC opposite to the less [XIX.]. Again, since AC is equal to AD, adding BA to both, we have the sum of the sides BA, AC equal to BD. *Therefore the sum of BA, AC is greater than BC.*

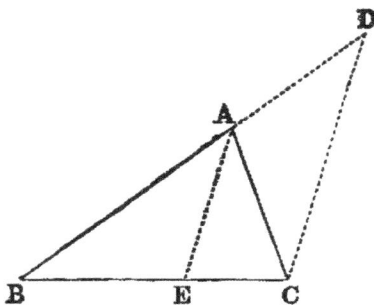

Or thus: Bisect the angle BAC by AE [IX.] Then the angle BEA is greater than EAC; but $EAC = EAB$ (const.); therefore the angle BEA is greater than EAB. Hence AB is greater than BE [XIX.]. In like manner AC is greater than EC. *Therefore the sum of BA, AC is greater than BC.*

Exercises.

1. In any triangle, the difference between any two sides is less than the third.

2. If any point within a triangle be joined to its angular points, the sum of the joining lines is greater than its semiperimeter.

3. If through the extremities of the base of a triangle, whose sides are unequal, lines be drawn to any point in the bisector of the vertical angle, their difference is less than the difference of the sides.

4. If the lines be drawn to any point in the bisector of the external vertical angle, their sum is greater than the sum of the sides.

5. Any side of any polygon is less than the sum of the remaining sides.

6. The perimeter of any triangle is greater than that of any inscribed triangle, and less than that of any circumscribed triangle.

7. The perimeter of any polygon is greater than that of any inscribed, and less than that of any circumscribed, polygon of the same number of sides.

8. The perimeter of a quadrilateral is greater than the sum of its diagonals.

Def.—*A line drawn from any angle of a triangle to the middle point of the opposite side is called a median of the triangle.*

9. The sum of the three medians of a triangle is less than its perimeter.

10. The sum of the diagonals of a quadrilateral is less than the sum of the lines which can be drawn to its angular points from any point except the intersection of the diagonals.

22

PROP. XXI.—Theorem.

If two lines (BD, CD) be drawn to a point (D) within a triangle from the extremities of its base (BC), their sum is less than the sum of the remaining sides (BA, CA), but they contain a greater angle.

Dem.—1. Produce BD (Post. II.) to meet AC in E. Then, in the triangle BAE, the sum of the sides BA, AE is greater than the side BE [XX.]: to each add EC, and we have the sum of BA, AC greater than the sum of BE, EC. Again, the sum of the sides DE, EC of the triangle DEC is greater than DC: to each add BD, and we get the sum of BE, EC greater than the sum of BD, DC; but it has been proved that the sum of BA, AC is greater than the sum of BE, EC. *Therefore much more is the sum of BA, AC greater than the sum of BD, DC.*

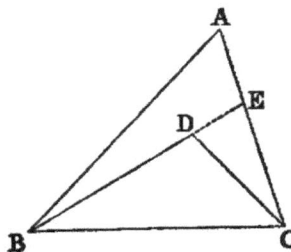

2. The external angle BDC of the triangle DEC is greater than the internal angle BEC [XVI.], and the angle BEC, for a like reason, is greater than BAC. *Therefore much more is BDC greater than BAC.*

Part 2 may be proved without producing either of the sides BD, DC. Thus: join AD and produce it to meet BC in F; then the angle BDF is greater than the angle BAF [XVI.], and FDC is greater than FAC. *Therefore the whole angle BDC is greater than BAC.*

Exercises.

1. The sum of the lines drawn from any point within a triangle to its angular points is less than the perimeter. (Compare Ex. 2, last Prop.)

2. If a convex polygonal line $ABCD$ lie within a convex polygonal line $AMND$ terminating in the same extremities, the length of the former is less than that of the latter.

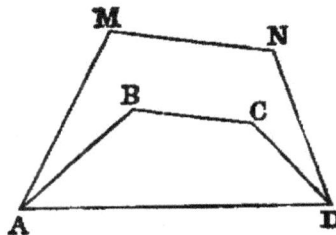

PROP. XXII.—Problem.

To construct a triangle whose three sides shall be respectively equal to three given lines (A, B, C), the sum of every two of which is greater than the third.

Sol.—Take any right line DE, terminated at D, but unlimited towards E, and cut off [III.] DF equal to A, FG equal to B, and GH equal to C. With F as centre, and FD as radius, describe the circle KDL (Post. III.); and with G as centre, and GH as radius, describe the circle KHL, intersecting the former circle in K. Join KF, KG. *KFG is the triangle required.*

23

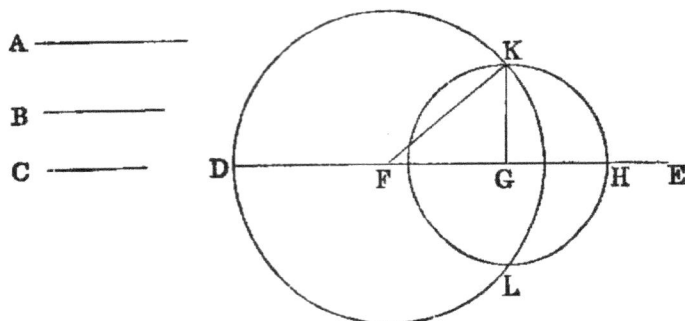

Dem.—Since F is the centre of the circle KDL, FK is equal to FD; but FD is equal to A (const.); therefore (Axiom I.) FK is equal to A. In like manner GK is equal to C, and FG is equal to B (const.) *Hence the three sides of the triangle KFG are respectively equal to the three lines A, B, C.*

Questions for Examination.

1. What is the reason for stating in the enunciation that the sum of every two of the given lines must be greater than the third?

2. Prove that when that condition is fulfilled the two circles must intersect.

3. Under what conditions would the circles not intersect?

4. If the sum of two of the lines were equal to the third, would the circles meet? Prove that they would not intersect.

PROP. XXIII.—Problem.

At a given point (A) in a given right line (AB) to make an angle equal to a given rectilineal angle (DEF).

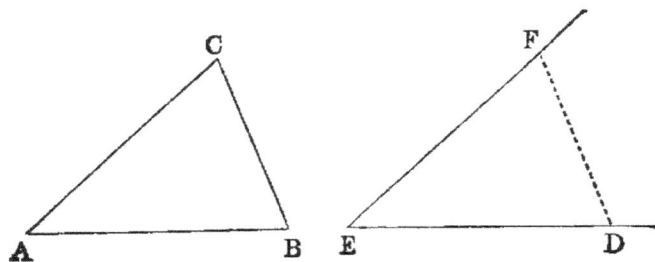

24

Sol.—In the sides *ED*, *EF* of the given angle take any arbitrary points *D* and *F*. Join *DF*, and construct [XXII.] the triangle *BAC*, whose sides, taken in order, shall be equal to those of *DEF*—namely, *AB* equal to *ED*, *AC* equal to *EF*, and *CB* equal to *FD*; then the angle *BAC* will [VIII.] be equal to *DEF*. *Hence it is the required angle.*

Exercises.

1. Construct a triangle, being given two sides and the angle between them.
2. Construct a triangle, being given two angles and the side between them.
3. Construct a triangle, being given two sides and the angle opposite to one of them.
4. Construct a triangle, being given the base, one of the angles at the base, and the sum or difference of the sides.
5. Given two points, one of which is in a given line, it is required to find another point in the given line, such that the sum or difference of its distances from the former points may be given. Show that two such points may be found in each case.

PROP. XXIV.—THEOREM.

If two triangles (ABC, DEF) have two sides (AB, AC) of one respectively equal to two sides (DE, DF) of the other, but the contained angle (BAC) of one greater than the contained angle (EDF) of the other, the base of that which has the greater angle is greater than the base of the other.

Dem.—Of the two sides *AB*, *AC*, let *AB* be the one which is not the greater, and with it make the angle *BAG* equal to *EDF* [XXIII.]. Then because *AB* is not greater than *AC*, *AG* is less than *AC* [XIX., Exer. 6]. Produce *AG* to *H*, and make *AH* equal to *DF* or *AC* [III.]. Join *BH*, *CH*.

In the triangles *BAH*, *EDF*, we have *AB* equal to *DE* (hyp.), *AH* equal to *DF* (const.), and the angle *BAH* equal to the angle *EDF* (const.); therefore the base [IV.] *BH* is equal to *EF*. Again, because *AH* is equal to *AC* (const.), the triangle *ACH* is isosceles; therefore the angle *ACH* is equal to *AHC* [V.]; but *ACH* is greater than *BCH*; therefore *AHC* is greater than *BCH*: much more is the angle *BHC* greater than

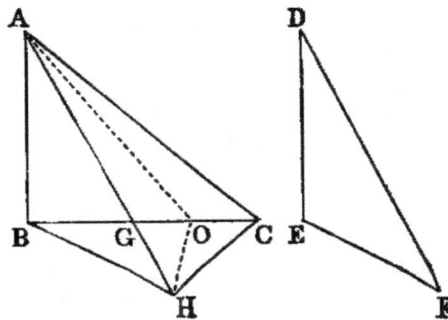

BCH, and the greater angle is subtended by the greater side [XIX.]. Therefore BC is greater than BH; but BH has been proved to be equal to EF; *therefore BC is greater than EF.*

The concluding part of this Proposition may be proved without joining CH, thus:—

$$BG + GH > BH \text{ [XX.]},$$
$$AG + GC > AC \text{ [XX.]};$$
therefore $\qquad BC + AH > BH + AC;$
but $\qquad AH = AC \text{ (const.)};$
therefore $\qquad\qquad BC \text{ is } > BH.$

Or thus: Bisect the angle CAH by AO. Join OH. Now in the △s CAO, HAO we have the sides CA, AO in one equal to the sides AH, AO in the other, and the contained angles equal; therefore the base OC is equal to the base OH [IV.]: to each add BO, and we have BC equal to the sum of BO, OH; but the sum of BO, OH is greater than BH [XX.]. *Therefore BC is greater than BH, that is, greater than EF.*

Exercises.

1. Prove this Proposition by making the angle ABH to the left of AB.
2. Prove that the angle BCA is greater than EFD.

PROP. XXV.—THEOREM.

If two triangles (ABC, DEF) have two sides (AB, AC) of one respectively equal to two sides (DE, DF) of the other, but the base (BC) of one greater than the base (EF) of the other, the angle (A) contained by the sides of that which has the greater base is greater them the angle (D) contained by the sides of the other.

Dem.—If the angle A be not greater than D, it must be either equal to it or less than it. We shall examine each case:—

1. If A were equal to D, the triangles ABC, DEF would have the two sides AB, AC of one respectively equal to the two sides DE, DF of the other, and the angle A contained by the two sides of one equal to the angle D contained by the two sides of the other. Hence [IV.] BC would be equal to EF; but BC is, by hypothesis, greater than EF; hence the angle A is not equal to the angle D.

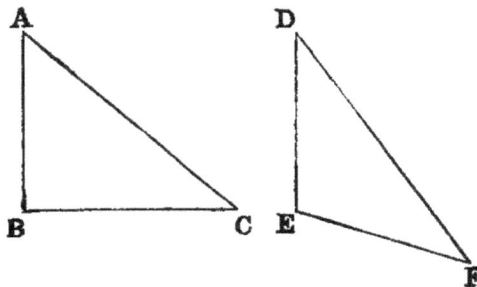

2. If A were less than D, then D would be greater than A, and the triangles DEF, ABC would have the two sides DE, DF of one respectively equal to the two sides AB, AC of the other, and the angle D contained by the two sides of one greater than the angle A contained by the two sides of the other. Hence [XXIV.] EF would be greater than BC; but EF (hyp.) is not greater than BC. Therefore A is not less than D, and we have proved that it is not equal to it; *therefore it must be greater.*

26

Or thus, directly: Construct the triangle ACG, whose three sides AG, GC, CA shall be respectively equal to the three sides DE, EF, FD of the triangle DEF [XXII.]. Join BG. Then because BC is greater than EF, BC is greater than CG. Hence [XVIII.] the angle BGC is greater than GBC; and make (XXIII.) the angle BGH equal to GBH, and join AH. Then [VI.] BH is equal to GH. Therefore the triangles ABH, AGH have the sides AB, AH of one equal to the sides AG, AH of the other, and the base BH equal to GH. Therefore [VIII.] the angle BAH is equal to GAH. *Hence the angle BAC is greater than CAG, and therefore greater than EDF.*

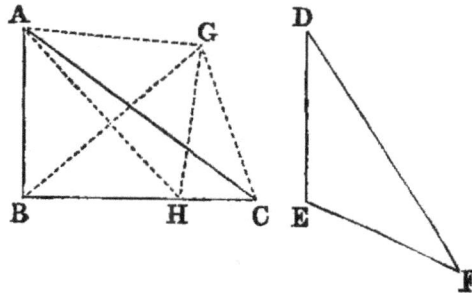

Exercise.

Demonstrate this Proposition directly by cutting off from BC a part equal to EF.

PROP. XXVI.—THEOREM.

If two triangles (ABC, DEF) have two angles (B, C) of one equal respectively to two angles (E, F) of the other, and a side of one equal to a side similarly placed with respect to the equal angles of the other, the triangles are equal in every respect.

Dem.—This Proposition breaks up into two according as the sides given to be equal are the sides adjacent to the equal angles, namely BC and EF, or those opposite equal angles.

1. Let the equal sides be BC and EF; then if DE be not equal to AB, suppose GE to be equal to it. Join GF; then the triangles ABC, GEF have the sides AB, BC of one respectively equal to the sides GE, EF of the other, and the angle ABC equal to the angle GEF (hyp.); therefore [IV.] the angle ACB is equal to the angle GFE; but the angle ACB is (hyp.) equal to DFE; hence GFE is equal to DFE—a part equal to the whole, which is absurd; therefore AB and DE are not unequal, that is, they are equal. Consequently the triangles ABC, DEF have the sides AB, BC of one respectively equal to the sides DE, EF of the other; and the contained angles ABC and DEF equal; *therefore* [IV.] *AC is equal to DF, and the angle BAC is equal to the angle EDF.*

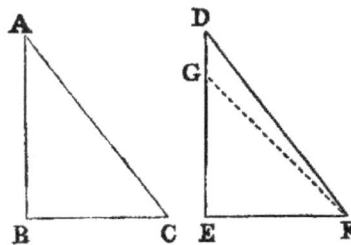

27

2. Let the sides given to be equal be AB and DE; it is required to prove that BC is equal to EF, and AC to DF. If BC be not equal to EF, suppose BG to be equal to it. Join AG. Then the triangles ABG, DEF have the two sides AB, BG of one respectively equal to the two sides DE, EF of the other, and the angle ABG equal to the angle DEF; therefore [IV.] the angle AGB is equal to DFE; but the angle ACB is equal to DFE (hyp.). Hence (Axiom I.) the angle AGB is equal to ACB, that is, the exterior angle of the triangle ACG is equal to the interior and non-adjacent angle, which [XVI.] is impossible. *Hence BC must be equal to EF, and the same as in 1, AC is equal to DF, and the angle BAC is equal to the angle EDF.*

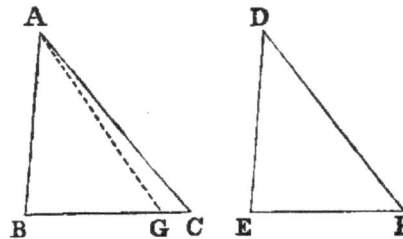

This Proposition, together with IV. and VIII., includes all the cases of the congruence of two triangles. Part I. may be proved immediately by superposition. For it is evident if ABC be applied to DEF, so that the point B shall coincide with E, and the line BC with EF, since BC is equal to EF, the point C shall coincide with F; and since the angles B, C are respectively equal to the angles E, F, the lines BA, CA shall coincide with ED and FD. *Hence the triangles are congruent.*

DEF.—*If every point on a geometrical figure satisfies an assigned condition, that figure is called the locus of the point satisfying the condition.* Thus, for example, a circle is the locus of a point whose distance from the centre is equal to its radius.

Exercises.

1. The extremities of the base of an isosceles triangle are equally distant from any point in the perpendicular from the vertical angle on the base.

2. If the line which bisects the vertical angle of a triangle also bisects the base, the triangle is isosceles.

3. The locus of a point which is equally distant from two fixed lines is the pair of lines which bisect the angles made by the fixed lines.

4. In a given right line find a point such that the perpendiculars from it on two given lines may be equal. State also the number of solutions.

5. If two right-angled triangles have equal hypotenuses, and an acute angle of one equal to an acute angle of the other, they are congruent.

6. If two right-angled triangles have equal hypotenuses, and a side of one equal to a side of the other, they are congruent.

7. The bisectors of the three internal angles of a triangle are concurrent.

8. The bisectors of two external angles and the bisector of the third internal angle are concurrent.

9. Through a given point draw a right line, such that perpendiculars on it from two given points on opposite sides may be equal to each other.

10. Through a given point draw a right line intersecting two given lines, and forming an isosceles triangle with them.

PARALLEL LINES.

28

Def. I.—*If two right lines in the same plane be such that, when produced indefinitely, they do not meet at any finite distance, they are said to be* PARALLEL.

Def. II.—A *parallelogram* is a quadrilateral, both pairs of whose opposite sides are parallel.

Def. III.—The right line joining either pair of opposite angles of a quadrilateral is called a *diagonal*.

Def. IV.—If both pairs of opposite sides of a quadrilateral be produced to meet, the right line joining their points of intersection is called its *third diagonal*.

Def. V.—A quadrilateral which has one pair of opposite sides parallel is called a *trapezium*.

Def. VI.—If from the extremities of one right line perpendiculars be drawn to another, the intercept between their feet is called the *projection* of the first line on the second.

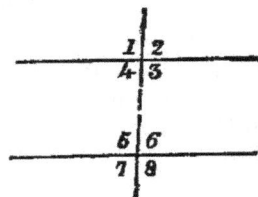

Def. VII.—When a right line intersects two other right lines in two distinct points it makes with them eight angles, which have received special names in relation to one another. Thus, in the figure—1, 2; 7, 8 are called *exterior* angles; 3, 4; 5, 6, *interior* angles. Again, 4; 6; 3, 5 are called *alternate* angles; lastly, 1, 5; 2, 6; 3, 8; 4, 7 are called *corresponding* angles.

PROP. XXVII.—Theorem.

If a right line (EF) intersecting two right lines (AB, CD) makes the alternate angles (AEF, EFD) equal to each other, these lines are parallel.

Dem.—If *AB* and *CD* are not parallel they must meet, if produced, at some finite distance: if possible let them meet in *G*; then the figure *EGF* is a triangle, and the angle *AEF* is an exterior angle, and *EFD* a non-adjacent interior angle. Hence [XVI.] *AEF* is greater than *EFD*; but it is also equal to it (hyp.), that is, both equal and greater, which is absurd. *Hence AB and CD are parallel.*

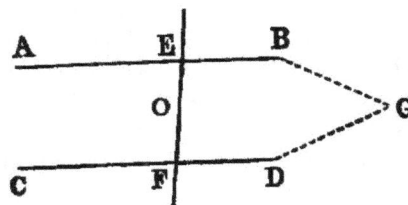

Or thus: Bisect *EF* in *O*; turn the whole figure round *O* as a centre, so that *EF* shall fall on itself; then because *OE* = *OF*, the point *E* shall fall on *F*; and because the angle *AEF* is equal to the angle *EFD*, the line *EA* will occupy the place of *FD*, and the line *FD* the place of *EA*; therefore the lines *AB*, *CD* interchange places, and the figure is symmetrical with respect to the point *O*. Hence, if *AB*, *CD* meet on one side of *O*, they must also meet on the other side; but two right lines cannot enclose a space (Axiom X.); therefore they do not meet at either side. *Hence they are parallel.*

PROP. XXVIII.—THEOREM.

If a right line (EF) intersecting two right lines (AB, CD) makes the exterior angle (EGB) equal to its corresponding interior angle (GHD), or makes two interior angles (BGH, GHD) on the same side equal to two right angles, the two right lines are parallel.

Dem.—1. Since the lines *AB, EF* intersect, the angle *AGH* is equal to *EGB* [XV.]; but *EGB* is equal to *GHD* (hyp.); therefore *AGH* is equal to *GHD*, and they are alternate angles. *Hence* [XXVII.] *AB is parallel to CD.*

2. Since *AGH* and *BGH* are adjacent angles, their sum is equal to two right angles [XIII.]; but the sum of *BGH* and *GHD* is two right angles (hyp.); therefore rejecting the angle *BGH* we have *AGH* equal *GHD*, and they are alternate angles; *therefore AB is parallel to CD* [XXVII.].

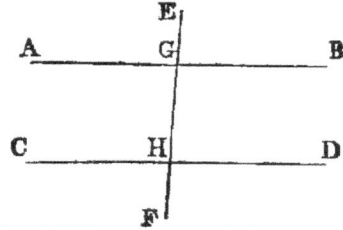

PROP. XXIX.—THEOREM.

If a right line (EF) intersect two parallel right lines (AB, CD), it makes— 1. the alternate angles (AGH, GHD) equal to one another; 2. the exterior angle (EGB) equal to the corresponding interior angle (GHD); 3. the two interior angles (BGH, GHD) on the same side equal to two right angles.

Dem.—If the angle *AGH* be not equal to *GHD*, one must be greater than the other. Let *AGH* be the greater; to each add *BGH*, and we have the sum of the angles *AGH, BGH* greater than the sum of the angles *BGH, GHD*; but the sum of *AGH, BGH* is two right angles; therefore the sum of *BGH, GHD* is less than two right angles, and therefore (Axiom XII.) the lines *AB, CD*, if produced, will meet at some finite distance: but since they are parallel (hyp.) they cannot meet at any finite distance. *Hence the angle AGH is not unequal to GHD—that is, it is equal to it.*

2. Since the angle *EGB* is equal to *AGH* [XV.], and *GHD* is equal to *AGH* (1), *EGB is equal to GHD* (Axiom I.).

3. Since *AGH* is equal to *GHD* (1), add *HGB* to each, and we have the sum of the angles *AGH, HGB* equal to the sum of the angles *GHD, HGB*; but the sum of the angles *AGH, HGB* [XIII.] is two right angles; *therefore the sum of the angles BGH, GHD is two right angles.*

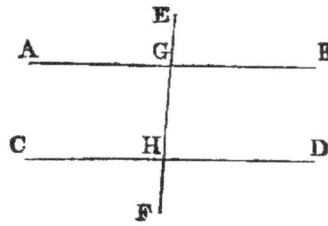

30

Exercises.

1. Demonstrate both parts of Prop. XXVIII. without using Prop. XXVII.

2. The parts of all perpendiculars to two parallel lines intercepted between them are equal.

3. If ACD, BCD be adjacent angles, any parallel to AB will meet the bisectors of these angles in points equally distant from where it meets CD.

4. If through the middle point O of any right line terminated by two parallel right lines any other secant be drawn, the intercept on this line made by the parallels is bisected in O.

5. Two right lines passing through a point equidistant from two parallels intercept equal portions on the parallels.

6. The perimeter of the parallelogram, formed by drawing parallels to two sides of an equilateral triangle from any point in the third side, is equal to twice the side.

7. If the opposite sides of a hexagon be equal and parallel, its diagonals are concurrent.

8. If two intersecting right lines be respectively parallel to two others, the angle between the former is equal to the angle between the latter. For if AB, AC be respectively parallel to DE, DF, and if AC, DE meet in G, the angles A, D are each equal to G [XXIX.].

PROP. XXX.—THEOREM.

If two right lines (AB, CD) be parallel to the same right line (EF), they are parallel to one another.

Dem.—Draw any secant GHK. Then since AB and EF are parallel, the angle AGH is equal to GHF [XXIX.]. In like manner the angle GHF is equal to HKD [XXIX.]. Therefore the angle AGK is equal to the angle GKD (Axiom I.). Hence [XXVII.] AB is parallel to CD.

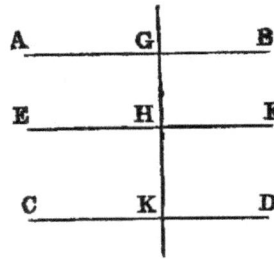

PROP. XXXI.—PROBLEM.

Through a given point (C) to draw a right line parallel to a given right line.

Sol.—Take any point D in AB. Join CD (Post. I.), and make the angle DCE equal to the angle ADC [XXIII.]. *The line CE is parallel to AB* [XXVII.].

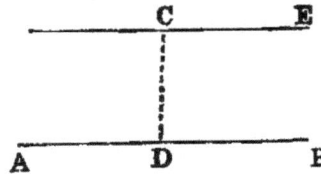

Exercises.

1. Given the altitude of a triangle and the base angles, construct it.

2. From a given point draw to a given line a line making with it an angle equal to a given angle. Show that there will be two solutions.

3. Prove the following construction for trisecting a given line AB:—On AB describe an equilateral $\triangle ABC$. Bisect the angles A, B by the lines AD, BD, meeting in D; through D draw parallels to AC, BC, meeting AB in E, F: E, F are the points of trisection of AB.

4. Inscribe a square in a given equilateral triangle, having its base on a given side of the triangle.

5. Draw a line parallel to the base of a triangle so that it may be—1. equal to the intercept it makes on one of the sides from the extremity of the base; 2. equal to the sum of the two intercepts on the sides from the extremities of the base; 3. equal to their difference. Show that there are two solutions in each case.

31

6. Through two given points in two parallel lines draw two lines forming a lozenge with the given parallels.

7. Between two lines given in position place a line of given length which shall be parallel to a given line. Show that there are two solutions.

PROP. XXXII.—THEOREM.

If any side (AB) of a triangle (ABC) be produced (to D), the external angle (CBD) is equal to the sum of the two internal non-adjacent angles (A, C), and the sum of the three internal angles is equal to two right angles.

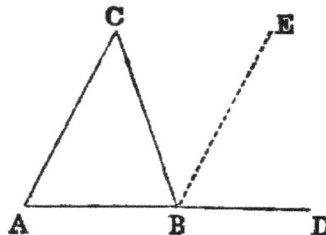

Dem.—Draw BE parallel to AC [XXXI.]. Now since BC intersects the parallels BE, AC, the alternate angles EBC, ACB are equal [XXIX.]. Again, since AB intersects the parallels BE, AC, the angle EBD is equal to BAC [XXIX.]; hence the whole angle CBD is equal to the sum of the two angles ACB, BAC: to each of these add the angle ABC and we have the sum of CBD, ABC equal to the sum of the three angles ACB, BAC, ABC: but the sum of CBD, ABC is two right angles [XIII.]; *hence the sum of the three angles ACB, BAC, ABC is two right angles.*

Cor. 1.—If a right-angled triangle be isosceles, each base angle is half a right angle.

Cor. 2.—If two triangles have two angles in one respectively equal to two angles in the other, their remaining angles are equal.

Cor. 3.—Since a quadrilateral can be divided into two triangles, the sum of its angles is equal to four right angles.

Cor. 4.—If a figure of n sides be divided into triangles by drawing diagonals from any one of its angles there will be $(n-2)$ triangles; hence the sum of its angles is equal $2(n-2)$ right angles.

Cor. 5.—If all the sides of any convex polygon be produced, the sum of the external angles is equal to four right angles.

Cor. 6.—Each angle of an equilateral triangle is two-thirds of a right angle.

Cor. 7.—If one angle of a triangle be equal to the sum of the other two, it is a right angle.

Cor. 8.—Every right-angled triangle can be divided into two isosceles triangles by a line drawn from the right angle to the hypotenuse.

Exercises.

1. Trisect a right angle.

2. Any angle of a triangle is obtuse, right, or acute, according as the opposite side is greater than, equal to, or less than, twice the *median* drawn from that angle.

3. If the sides of a polygon of n sides be produced, the sum of the angles between each alternate pair is equal to $2(n-4)$ right angles.

32

4. If the line which bisects the external vertical angle be parallel to the base, the triangle is isosceles.

5. If two right-angled △s ABC, ABD be on the same hypotenuse AB, and the vertices C and D be joined, the pair of angles subtended by any side of the quadrilateral thus formed are equal.

6. The three perpendiculars of a triangle are concurrent.

7. The bisectors of two adjacent angles of a parallelogram are at right angles.

8. The bisectors of the external angles of a quadrilateral form a circumscribed quadrilateral, the sum of whose opposite angles is equal to two right angles.

9. If the three sides of one triangle be respectively perpendicular to those of another triangle, the triangles are equiangular.

10. Construct a right-angled triangle, being given the hypotenuse and the sum or difference of the sides.

11. The angles made with the base of an isosceles triangle by perpendiculars from its extremities on the equal sides are each equal to half the vertical angle.

12. The angle included between the internal bisector of one base angle of a triangle and the external bisector of the other base angle is equal to half the vertical angle.

13. In the construction of Prop. XVIII. prove that the angle DBC is equal to half the difference of the base angles.

14. If A, B, C denote the angles of a △, prove that $\frac{1}{2}(A + B)$, $\frac{1}{2}(B + C)$, $\frac{1}{2}(C + A)$ will be the angles of a △ formed by any side and the bisectors of the external angles between that side and the other sides produced.

PROP. XXXIII.—THEOREM.

The right lines (AC, BD) which join the adjacent extremities of two equal and parallel right lines (AB, CD) are equal and parallel.

Dem.—Join BC. Now since AB is parallel to CD, and BC intersects them, the angle ABC is equal to the alternate angle DCB [XXIX.]. Again, since AB is equal to CD, and BC common, the triangles ABC, DCB have the sides AB, BC in one respectively equal to the sides DC, CB in the other, and the angles ABC, DCB contained by those sides equal; therefore [IV.] the base AC is equal to the base BD, and the angle ACB is equal to the angle CBD; but these are alternate angles; hence [XXVII.] AC is parallel to BD, and it has been proved equal to it. *Therefore AC is both equal and parallel to BD.*

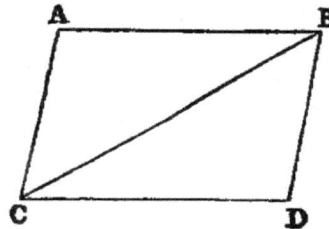

Exercises.

1. If two right lines AB, BC be respectively equal and parallel to two other right lines DE, EF, the right line AC joining the extremities of the former pair is equal to the right line DF joining the extremities of the latter.

2. Right lines that are equal and parallel have equal projections on any other right line; and conversely, parallel right lines that have equal projections on another right line are equal.

3. Equal right lines that have equal projections on another right line are parallel.

4. The right lines which join transversely the extremities of two equal and parallel right lines bisect each other.

PROP. XXXIV.—THEOREM.

The opposite sides (AB, CD; AC, BD) and the opposite angles (A, D; B, C) of a parallelogram are equal to one another, and either diagonal bisects the parallelogram.

Dem.—Join BC. Since AB is parallel to CD, and BC intersects them, the angle ABC is equal to the angle BCD [XXIX.]. Again, since BC intersects the parallels AC, BD, the angle ACB is equal to the angle CBD; hence the triangles ABC, DCB have the two angles ABC, ACB in one respectively equal to the two angles BCD, CBD in the other, and the side BC common. *Therefore* [XXVI.] *AB is equal to CD, and AC to BD; the angle BAC to the angle BDC, and the triangle ABC to the triangle BDC.*

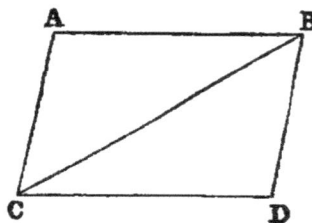

Again, because the angle ACB is equal to CBD, and DCB equal to ABC, *the whole angle ACD is equal to the whole angle ABD.*

Cor. 1.—If one angle of a parallelogram be a right angle, all its angles are right angles.

Cor. 2.—If two adjacent sides of a parallelogram be equal, it is a lozenge.

Cor. 3.—If both pairs of opposite sides of a quadrilateral be equal, it is a parallelogram.

Cor. 4.—If both pairs of opposite angles of a quadrilateral be equal, it is a parallelogram.

Cor. 5.—If the diagonals of a quadrilateral bisect each other, it is a parallelogram.

Cor. 6.—If both diagonals of a quadrilateral bisect the quadrilateral, it is a parallelogram.

Cor. 7.—If the adjacent sides of a parallelogram be equal, its diagonals bisect its angles.

Cor. 8.—If the adjacent sides of a parallelogram be equal, its diagonals intersect at right angles.

Cor. 9.—In a right-angled parallelogram the diagonals are equal.

Cor. 10.—If the diagonals of a parallelogram be perpendicular to each other, it is a lozenge.

Cor. 11.—If a diagonal of a parallelogram bisect the angles whose vertices it joins, the parallelogram is a lozenge.

Exercises.

1. The diagonals of a parallelogram bisect each other.
2. If the diagonals of a parallelogram be equal, all its angles are right angles.
3. Divide a right line into any number of equal parts.
4. The right lines joining the adjacent extremities of two unequal parallel right lines will meet, if produced, on the side of the shorter parallel.
5. If two opposite sides of a quadrilateral be parallel but not equal, and the other pair equal but not parallel, its opposite angles are supplemental.
6. Construct a triangle, being given the middle points of its three sides.

34

7. The area of a quadrilateral is equal to the area of a triangle, having two sides equal to its diagonals, and the contained angle equal to that between the diagonals.

PROP. XXXV.—THEOREM.

Parallelograms on the same base (BC) and between the same parallels are equal.

Dem.—1. Let the sides AD, DF of the parallelograms AC, BF opposite to the common base BC terminate in the same point D, then [XXXIV.] each parallelogram is double of the triangle BCD. *Hence they are equal to one another.*

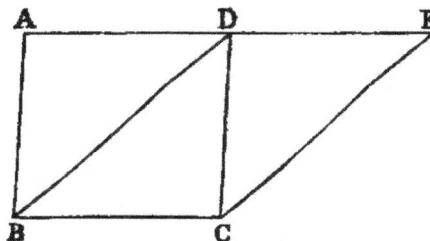

2. Let the sides AD, EF (figures (α), (β)) opposite to BC not terminate in the same point.

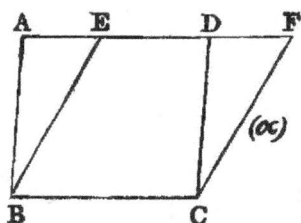

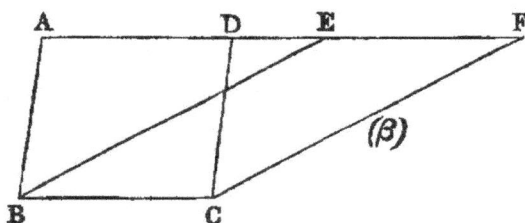

Then because $ABCD$ is a parallelogram, AD is equal to BC [XXXIV.]; and since $BCEF$ is a parallelogram, EF is equal to BC; therefore (see fig. (α)) take away ED, and in fig. (β) add ED, and we have in each case AE equal to DF, and BA is equal to CD [XXXIV.]. Hence the triangles BAE, CDF have the two sides BA, AE in one respectively equal to the two sides CD, DF in the other, and the angle BAE [XXIX.] equal to the angle CDF; hence [IV.] the triangle BAE is equal to the triangle CDF; and taking each of these triangles in succession from the quadrilateral $BAFC$, *there will remain the parallelogram $BCFE$ equal to the parallelogram $BCDA$.*

Or thus: The triangles ABE, DCF have [XXXIV.] the sides AB, BE in one respectively equal to the sides DC, CF in the other, and the angle ABE equal to the angle DCF [XXIX., Ex. 8]. Hence the triangle ABE is equal to the triangle DCF; and, taking each away from the quadrilateral $BAFC$, *there will remain the parallelogram $BCFE$ equal to the parallelogram $BCDA$.*

Observation.—By the second method of proof the subdivision of the demonstration into cases is avoided. It is easy to see that either of the two parallelograms $ABCD$, $EBCF$ can be divided into parts and rearranged so as to make it congruent with the other. This Proposition affords the first instance in the Elements in which equality which is not congruence occurs. This equality is expressed algebraically by the symbol $=$, while congruence is denoted by $\equiv$, called also the symbol of identity. Figures that are congruent are said to be *identically equal*.

35

PROP. XXXVI.—Theorem.

Parallelograms (BD, FH) on equal bases (BC, FG) and between the same parallels are equal.

Dem.—Join BE, CH. Now since FH is a parallelogram, FG is equal to EH [XXXIV.]; but BC is equal to FG (hyp.); therefore BC is equal to EH (Axiom I.). Hence BE, CH, which join their adjacent extremities, are equal and parallel; therefore BH is a parallelogram. Again, since the parallelograms BD, BH are on the same base BC, and between the same parallels BC, AH, they are equal [XXXV.]. In like manner, since the parallelograms HB, HF are on the same base EH, and between the same parallels EH, BG, they are equal. Hence BD and FH are each equal to BH. *Therefore* (Axiom I.) *BD is equal to FH.*

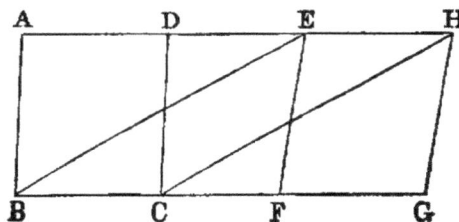

Exercise.—Prove this Proposition without joining BE, CH.

PROP. XXXVII.—Theorem.

Triangles (ABC, DBC) on the same base (BC) and between the same parallels (AD, BC) are equal.

Dem.—Produce AD both ways. Draw BE parallel to AC, and CF parallel to BD [XXXI.] Then the figures $AEBC$, $DBCF$ are parallelograms; and since they are on the same base BC, and between the same parallels BC, EF they are equal [XXXV.]. Again, the triangle ABC is half the parallelogram $AEBC$ [XXXIV.], because the diagonal AB bisects it. In like manner the triangle DBC is half the parallelogram $DBCF$, because the diagonal DC bisects it, and halves of equal things are equal (Axiom VII.). *Therefore the triangle ABC is equal to the triangle DBC.*

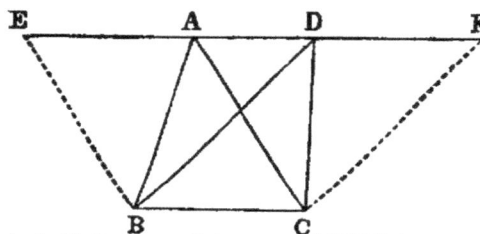

Exercises.

1. If two equal triangles be on the same base, but on opposite sides, the right line joining their vertices is bisected by the base.

2. Construct a triangle equal in area to a given quadrilateral figure.

3. Construct a triangle equal in area to a given rectilineal figure.

4. Construct a lozenge equal to a given parallelogram, and having a given side of the parallelogram for base.

5. Given the base and the area of a triangle, find the locus of the vertex.

6. If through a point O, in the production of the diagonal AC of a parallelogram $ABCD$, any right line be drawn cutting the sides AB, BC in the points E, F, and ED, FD be joined, the triangle EFD is less than half the parallelogram.

PROP. XXXVIII.—Theorem.

Two triangles on equal bases and between the same parallels are equal.

Dem.—By a construction similar to the last, we see that the triangles are the halves of parallelograms, on equal bases, and between the same parallels. Hence they are the halves of equal parallelograms [XXXVI.]. *Therefore they are equal to one another.*

Exercises.

1. Every median of a triangle bisects the triangle.
2. If two triangles have two sides of one respectively equal to two sides of the other, and the contained angles supplemental, their areas are equal.
3. If the base of a triangle be divided into any number of equal parts, right lines drawn from the vertex to the points of division will divide the whole triangle into as many equal parts.
4. Right lines from any point in the diagonal of a parallelogram to the angular points through which the diagonal does not pass, and the diagonal, divide the parallelogram into four triangles which are equal, two by two.
5. If one diagonal of a quadrilateral bisects the other, it also bisects the quadrilateral, and conversely.
6. If two △s ABC, ABD be on the same base AB, and between the same parallels, and if a parallel to AB meet the sides AC, BC in the point E, F; and the sides AD, BD in the point G, H; then $EF = GH$.
7. If instead of triangles on the same base we have triangles on equal bases and between the same parallels, the intercepts made by the sides of the triangles on any parallel to the bases are equal.
8. If the middle points of any two sides of a triangle be joined, the triangle so formed with the two half sides is one-fourth of the whole.
9. The triangle whose vertices are the middle points of two sides, and any point in the base of another triangle, is one-fourth of that triangle.
10. Bisect a given triangle by a right line drawn from a given point in one of the sides.
11. Trisect a given triangle by three right lines drawn from a given point within it.
12. Prove that any right line through the intersection of the diagonals of a parallelogram bisects the parallelogram.
13. The triangle formed by joining the middle point of one of the non-parallel sides of a trapezium to the extremities of the opposite side is equal to half the trapezium.

PROP. XXXIX.—Theorem.

Equal triangles (BAC, BDC) on the same base (BC) and on the same side of it are between the same parallels.

Dem.—Join AD. Then if AD be not parallel to BC, let AE be parallel to it, and let it cut BD in E. Join EC. Now since the triangles BEC, BAC are on the same base BC, and between the same parallels BC, AE, they are equal [XXXVII.]; but the triangle BAC is equal to the triangle BDC (hyp.). Therefore (Axiom I.) the triangle BEC is equal to the triangle BDC—that is, a part equal to the whole which is absurd. *Hence AD must be parallel to BC.*

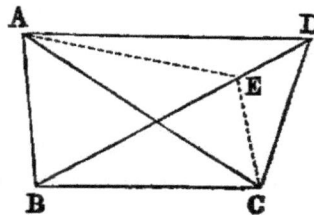

37

PROP. XL.—Theorem.

Equal triangles (ABC, DEF) on equal bases (BC, EF) which form parts of the same right line, and on the same side of the line, are between the same parallels.

Dem.—Join *AD*. If *AD* be not parallel to *BF*, let *AG* be parallel to it. Join *GF*. Now since the triangles *GEF* and *ABC* are on equal bases *BC*, *EF*, and between the same parallels *BF*, *AG*, they are equal [XXXVIII.]; but the triangle *DEF* is equal to the triangle *ABC* (hyp.). Hence *GEF* is equal to *DEF* (Axiom I.)— that is, a part equal to the whole, which is absurd. *Therefore AD must be parallel to BF.*

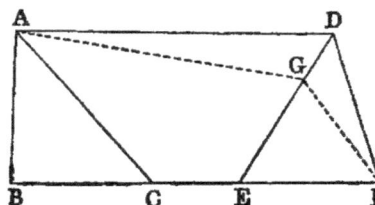

DEF.—*The altitude of a triangle is the perpendicular from the vertex on the base.*

Exercises.

1. Triangles and parallelograms of equal bases and altitudes are respectively equal.

2. The right line joining the middle points of two sides of a triangle is parallel to the third; for the medians from the extremities of the base to these points will each bisect the original triangle. Hence the two triangles whose base is the third side and whose vertices are the points of bisection are equal.

3. The parallel to any side of a triangle through the middle point of another bisects the third.

4. The lines of connexion of the middle points of the sides of a triangle divide it into four congruent triangles.

5. The line of connexion of the middle points of two sides of a triangle is equal to half the third side.

6. The middle points of the four sides of a convex quadrilateral, taken in order, are the angular points of a parallelogram whose area is equal to half the area of the quadrilateral.

7. The sum of the two parallel sides of a trapezium is double the line joining the middle points of the two remaining sides.

8. The parallelogram formed by the line of connexion of the middle points of two sides of a triangle, and any pair of parallels drawn through the same points to meet the third side, is equal to half the triangle.

9. The right line joining the middle points of opposite sides of a quadrilateral, and the right line joining the middle points of its diagonals, are concurrent.

PROP. XLI.—Theorem.

If a parallelogram (ABCD) and a triangle (EBC) be on the same base (BC) and between the same parallels, the parallelogram is double of the triangle.

38

Dem.—Join *AC*.   The parallelogram *ABCD* is double of the triangle *ABC* [XXXIV.]; but the triangle *ABC* is equal to the triangle *EBC* [XXXVII.]. *Therefore the parallelogram ABCD is double of the triangle EBC.*

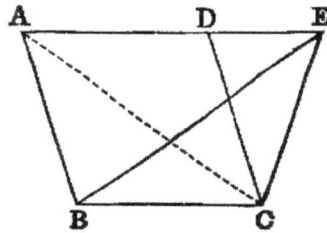

Cor. 1.—If a triangle and a parallelogram have equal altitudes, and if the base of the triangle be double of the base of the parallelogram, the areas are equal.

Cor. 2.—The sum of the triangles whose bases are two opposite sides of a parallelogram, and which have any point between these sides as a common vertex, is equal to half the parallelogram.

PROP. XLII.—PROBLEM.

To construct a parallelogram equal to a given triangle (ABC), and having an angle equal to a given angle (D).

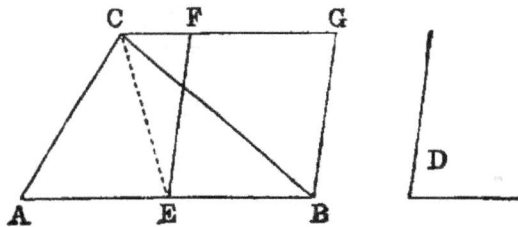

Sol.—Bisect *AB* in *E*. Join *EC*. Make the angle *BEF* [XXIII.] equal to *D*. Draw *CG* parallel to *AB* [XXXI.], and *BG* parallel to *EF*. *EG is a parallelogram fulfilling the required conditions.*

Dem.—Because *AE* is equal to *EB* (const.), the triangle *AEC* is equal to the triangle *EBC* [XXXVIII.], therefore the triangle *ABC* is double of the triangle *EBC*; but the parallelogram *EG* is also double of the triangle *EBC* [XLI.], because they are on the same base *EB*, and between the same parallels *EB* and *CG*. Therefore the parallelogram *EG* is equal to the triangle *ABC*, and it has (const.) the angle *BEF* equal to *D*. *Hence EG is a parallelogram fulfilling the required conditions.*

PROP. XLIII.—THEOREM.

The parallels (EF, GH) through any point (K) in one of the diagonals (AC) of a parallelogram divide it into four parallelograms, of which the two (BK, KD) through which the diagonal does not pass, and which

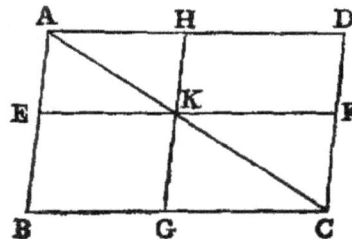

39

are called the COMPLEMENTS *of the other two, are equal.*

Dem.—Because the diagonal bisects the parallelograms AC, AK, KC we have [XXXIV.] the triangle ADC equal to the triangle ABC, the triangle AHK equal to AEK, and the triangle KFC equal to the triangle KGC. Hence, subtracting the sums of the two last equalities from the first, we get *the parallelogram DK equal to the parallelogram KB.*

Cor. 1.—If through a point K within a parallelogram $ABCD$ lines drawn parallel to the sides make the parallelograms DK, KB equal, K is a point in the diagonal AC.

Cor. 2.—The parallelogram BH is equal to AF, and BF to HC.

Cor. 2. supplies an easy demonstration of a fundamental Proposition in Statics.

Exercises.

1. *If EF, GH be parallels to the adjacent sides of a parallelogram ABCD, the diagonals EH, GF of two of the four* □*s into which they divide it and one of the diagonals of ABCD are concurrent.*

Dem.—Let EH, GF meet in M; through M draw MP, MJ parallel to AB, BC. Produce AD, GH, BC to meet MP, and AB, EF, DC to meet MJ. Now the complement $OF = FJ$: to each add the □ FL, and we get the figure $OFL = $ □ CJ. Again, the complement $PH = HK$ [XLIII.]: to each add the □ OC, and we get the □ $PC = $ figure OFL. Hence the □ $PC = CJ$. Therefore they are about the same diagonal [XLIII., *Cor.* 1]. *Hence AC produced will pass through M.*

2. *The middle points of the three diagonals AC, BD, EF of a quadrilateral ABCD are collinear.*

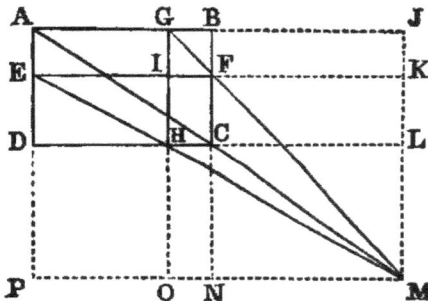

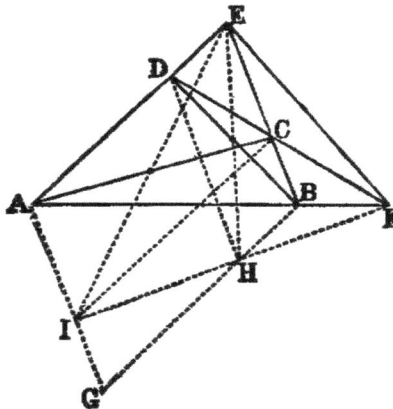

Dem.—Complete the □ $AEBG$. Draw DH, CI parallel to AG, BG. Join IH, and produce; then AB, CD, IH are concurrent (Ex. 1); therefore IH will pass through F. Join

40

EI, EH. Now [XI., Ex. 2, 3] *the middle points of EI, EH, EF are collinear, but* [XXXIV., Ex. 1] *the middle points of EI, EH are the middle points of AC, BD. Hence the middle points of AC, BD, EF are collinear.*

PROP. XLIV.—PROBLEM.

To a given, right line (AB) to apply a parallelogram which shall be equal to a given triangle (C), and have one of its angles equal to a given angle (D).

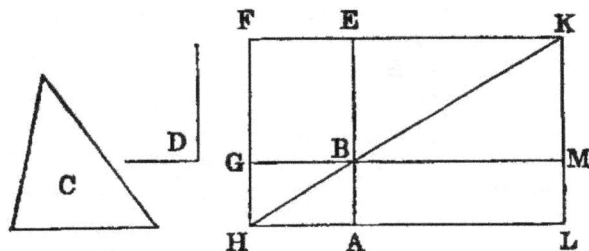

Sol.—Construct the parallelogram *BEFG* [XLII.] equal to the given triangle *C*, and having the angle *B* equal to the given angle *D*, and so that its side *BE* shall be in the same right line with *AB*. Through *A* draw *AH* parallel to *BG* [XXXI.], and produce *FG* to meet it in *H*. Join *HB*. Then because *HA* and *FE* are parallels, and *HF* intersects them, the sum of the angles *AHF, HFE* is two right angles [XXIX.]; therefore the sum of the angles *BHF, HFE* is less than two right angles; and therefore (Axiom XII.) the lines *HB, FE*, if produced, will meet as at *K*. Through *K* draw *KL* parallel to *AB* [XXXI.], and produce *HA* and *GB* to meet it in the points *L* and *M. Then AM is a parallelogram fulfilling the required conditions.*

Dem.—The parallelogram *AM* is equal to *GE* [XLIII.]; but *GE* is equal to the triangle *C* (const.); therefore *AM* is equal to the triangle *C*. Again, the angle *ABM* is equal to *EBG* [XV.], and *EBG* is equal to *D* (const.); therefore the angle *ABM* is equal to *D*; and *AM* is constructed on the given line; *therefore it is the parallelogram required.*

PROP. XLV.—PROBLEM.

To construct a parallelogram equal to a given rectilineal figure (ABCD), and having an angle equal to a given rectilineal angle (X).

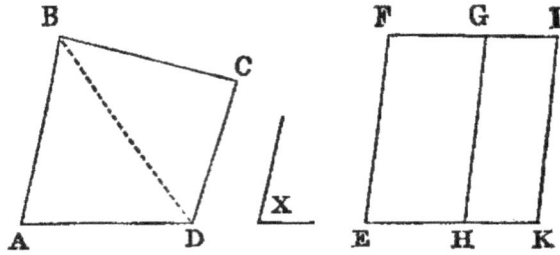

Sol.—Join BD. Construct a parallelogram EG [XLII.] equal to the triangle ABD, and having the angle E equal to the given angle X; and to the right line GH apply the parallelogram HI equal to the triangle BCD, and having the angle GHK equal to X [XLIV.], and so on for additional triangles if there be any. *Then EI is a parallelogram fulfilling the required conditions.*

Dem.—Because the angles GHK, FEH are each equal to X (const.), they are equal to one another: to each add the angle GHE, and we have the sum of the angles GHK, GHE equal to the sum of the angles FEH, GHE; but since HG is parallel to EF, and EH intersects them, the sum of FEH, GHE is two right angles [XXIX.]. Hence the sum of GHK, GHE is two right angles; therefore EH, HK are in the same right line [XIV.].

Again, because GH intersects the parallels FG, EK, the alternate angles FGH, GHK are equal [XXIX.]: to each add the angle HGI, and we have the sum of the angles FGH, HGI equal to the sum of the angles GHK, HGI; but since GI is parallel to HK, and GH intersects them, the sum of the angles GHK, HGI is equal to two right angles [XXIX.]. Hence the sum of the angles FGH, HGI is two right angles; therefore FG and GI are in the same right line [XIV.].

Again, because EG and HI are parallelograms, EF and KI are each parallel to GH; hence [XXX.] EF is parallel to KI, and the opposite sides EK and FI are parallel; therefore EI is a parallelogram; and because the parallelogram EG (const.) is equal to the triangle ABD, and HI to the triangle BCD, the whole parallelogram EI is equal to the rectilineal figure $ABCD$, and it has the angle E equal to the given angle X. *Hence EI is a parallelogram fulfilling the required conditions.*

It would simplify Problems XLIV., XLV., if they were stated as the constructing of rectangles, and in this special form they would be better understood by the student, since rectangles are the simplest areas to which others are referred.

Exercises.

1. Construct a rectangle equal to the sum of two or any number of rectilineal figures.
2. Construct a rectangle equal to the difference of two given figures.

PROP. XLVI.—PROBLEM.

On a given right line (AB) to describe a square.

Sol.—Erect AD at right angles to AB [XI.], and make it equal to AB [III.]. Through D draw DC parallel to AB [XXXI.], and through B draw BC parallel to AD; *then AC is the square required.*

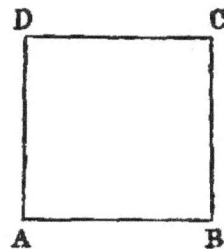

Dem.—Because AC is a parallelogram, AB is equal to CD [XXXIV.]; but AB is equal to AD (const.); therefore AD is equal to CD, and AD is equal to BC [XXXIV.]. Hence the four sides are equal; therefore AC is a lozenge, and the angle A is a right angle. *Therefore AC is a square* (Def. XXX.).

Exercises.

1. The squares on equal lines are equal; and, conversely, the sides of equal squares are equal.

2. The parallelograms about the diagonal of a square are squares.

3. If on the four sides of a square, or on the sides produced, points be taken equidistant from the four angles, they will be the angular points of another square, and similarly for a regular pentagon, hexagon, &c.

4. Divide a given square into five equal parts; namely, four right-angled triangles, and a square.

PROP. XLVII.—THEOREM.

In a right-angled triangle (ABC) the square on the hypotenuse (AB) is equal to the sum of the squares on the other two sides (AC, BC).

Dem.—On the sides AB, BC, CA describe squares [XLVI.]. Draw CL parallel to AG. Join CG, BK. Then because the angle ACB is right (hyp.), and ACH is right, being the angle of a square, the sum of the angles ACB, ACH is two right angles; therefore BC, CH are in the same right line [XIV.]. In like manner AC, CD are in the same right line. Again, because BAG is the angle of a square it is a right angle: in like manner CAK is a right angle. Hence BAG is equal to CAK: to each add BAC, and we get the angle CAG equal to KAB. Again, since BG and CK are squares, BA is equal to AG, and CA to AK. Hence the two triangles CAG, KAB have the sides CA, AG in one respectively equal to the sides KA, AB in the other, and the contained angles CAG,

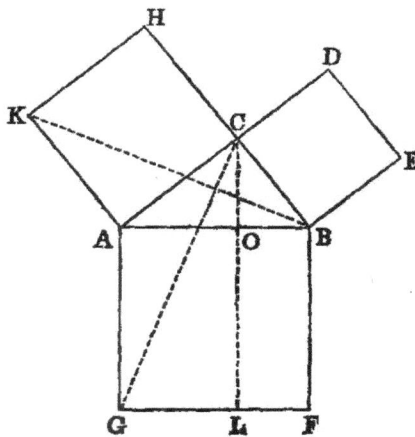

43

KAB also equal. Therefore [IV.] the triangles are equal; but the parallelogram AL is double of the triangle CAG [XLI.], because they are on the same base AG, and between the same parallels AG and CL. In like manner the parallelogram AH is double of the triangle KAB, because they are on the same base AK, and between the same parallels AK and BH; and since doubles of equal things are equal (Axiom VI.), the parallelogram AL is equal to AH. In like manner it can be proved that the parallelogram BL is equal to BD. *Hence the whole square AF is equal to the sum of the two squares AH and BD.*

Or thus: Let all the squares be made in reversed directions. Join CG, BK, and through C draw OL parallel to AG. Now, taking the $\angle BAC$ from the right $\angle$s BAG, CAK, the remaining $\angle$s CAG, BAK are equal. Hence the $\triangle$s CAG, BAK have the side $CA = AK$, and $AG = AB$, and the $\angle CAG = BAK$; therefore [IV.] they are equal; and since [XLI.] the $\square$s AL, AH are respectively the doubles of these triangles, they are equal. In like manner the $\square$s BL, BD are equal; hence the whole square AF is equal to the sum of the two squares AH, BD.

This proof is shorter than the usual one, since it is not necessary to prove that AC, CD are in one right line. In a similar way the Proposition may be proved by taking any of the eight figures formed by turning the squares in all possible directions. Another simplification of the proof would be got by considering that the point A is such that one of the $\triangle$s CAG, BAK can be turned round it in its own plane until it coincides with the other; and hence that they are congruent.

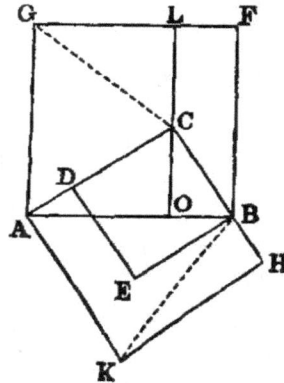

Exercises.

1. The square on AC is equal to the rectangle $AB \cdot AO$, and the square on $BC = AB \cdot BO$.
2. The square on $CO = AO \cdot OB$.
3. $AC^2 - BC^2 = AO^2 - BO^2$.
4. Find a line whose square shall be equal to the sum of two given squares.
5. Given the base of a triangle and the difference of the squares of its sides, the locus of its vertex is a right line perpendicular to the base.
6. The transverse lines BK, CG are perpendicular to each other.
7. If EG be joined, its square is equal to $AC^2 + 4BC^2$.
8. The square described on the sum of the sides of a right-angled triangle exceeds the square on the hypotenuse by four times the area of the triangle (*see* fig., XLVI., Ex. 3). More generally, if the vertical angle of a triangle be equal to the angle of a regular polygon of n sides, then the regular polygon of n sides, described on a line equal to the sum of its sides, exceeds the area of the regular polygon of n sides described on the base by n times the area of the triangle.
9. If AC and BK intersect in P, and through P a line be drawn parallel to BC, meeting AB in Q; then CP is equal to PQ.
10. Each of the triangles AGK and BEF, formed by joining adjacent corners of the squares, is equal to the right-angled triangle ABC.
11. Find a line whose square shall be equal to the difference of the squares on two lines.
12. The square on the difference of the sides AC, CB is less than the square on the hypotenuse by four times the area of the triangle.
13. If AE be joined, the lines AE, BK, CL, are concurrent.
14. In an equilateral triangle, three times the square on any side is equal to four times the square on the perpendicular to it from the opposite vertex.
15. On BE, a part of the side BC of a square $ABCD$, is described the square $BEFG$, having its side BG in the continuation of AB; it is required to divide the figure $AGFECD$ into three parts which will form a square.
16. Four times the sum of the squares on the medians which bisect the sides of a right-angled triangle is equal to five times the square on the hypotenuse.
17. If perpendiculars be let fall on the sides of a polygon from any point, dividing each side into two segments, the sum of the squares on one set of alternate segments is equal to the sum of the squares on the remaining set.
18. The sum of the squares on lines drawn from any point to one pair of opposite angles of a rectangle is equal to the sum of the squares on the lines from the same point to the remaining pair.
19. Divide the hypotenuse of a right-angled triangle into two parts, such that the difference between their squares shall be equal to the square on one of the sides.

20. From the extremities of the base of a triangle perpendiculars are let fall on the opposite sides; prove that the sum of the rectangles contained by the sides and their lower segments is equal to the square on the base.

PROP. XLVIII.—THEOREM.

If the square on one side (AB) of a triangle be equal to the sum of the squares on the remaining sides (AC, CB), the angle (C) opposite to that side is a right angle.

Dem.—Erect CD at right angles to CB [XI.], and make CD equal to CA [III.]. Join BD. Then because AC is equal to CD, the square on AC is equal to the square on CD: to each add the square on CB, and we have the sum of the squares on AC, CB equal to the sum of the squares on CD, CB; but the sum of the squares on AC, CB is equal to the square on AB (hyp.), and the sum of the squares on CD, CB is equal to the square on BD [XLVII.]. Therefore the square on AB is equal to the square on BD. Hence AB is equal to BD [XLVI., Ex. 1]. Again, because AC is equal to CD (const.), and CB common to the two triangles ACB, DCB, and the base AB equal to the base DB, the angle ACB is equal to the angle DCB; but the angle DCB is a right angle (const.). *Hence the angle ACB is a right angle.*

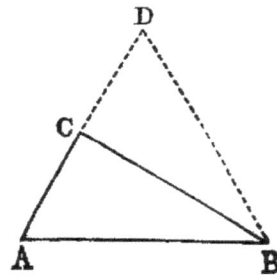

The foregoing proof forms an exception to Euclid's demonstrations of converse propositions, for it is direct. The following is an indirect proof:—If CB be not at right angles to AC, let CD be perpendicular to it. Make $CD = CB$. Join AD. Then, as before, it can be proved that AD is equal to AB, and CD is equal to CB (const.). This is contrary to Prop. VII. *Hence the angle ACB is a right angle.*

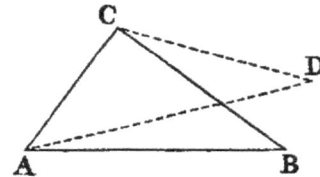

Questions for Examination on Book I.

1. What is Geometry?
2. What is geometric magnitude? *Ans.* That which has extension in space.
3. Name the primary concepts of geometry. *Ans.* Points, lines, surfaces, and solids.
4. How may lines be divided? *Ans.* Into straight and curved.
5. How is a straight line generated? *Ans.* By the motion of a point which has the same direction throughout.
6. How is a curved line generated? *Ans.* By the motion of a point which continually changes its direction.
7. How may surfaces be divided? *Ans.* Into planes and curved surfaces.
8. How may a plane surface be generated. *Ans.* By the motion of a right line which crosses another right line, and moves along it without changing its direction.
9. Why has a point no dimensions?
10. Why has a line neither breadth nor thickness?
11. How many dimensions has a surface?
12. What is Plane Geometry?

46

13. What portion of plane geometry forms the subject of the "First Six Books of Euclid's Elements"? *Ans.* The geometry of the *point, line,* and *circle.*

14. What is the subject-matter of Book I.?

15. How many conditions are necessary to fix the position of a point in a plane? *Ans.* Two; for it must be the intersection of two lines, straight or curved.

16. Give examples taken from Book I.

17. In order to construct a line, how many conditions must be given? *Ans.* Two; as, for instance, two points through which it must pass; or one point through which it must pass and a line to which it must be parallel or perpendicular, &c.

18. What problems on the drawing of lines occur in Book I.? *Ans.* II., IX., XI., XII., XXIII., XXXI., in each of which, except Problem 2, there are two conditions. The direction in Problem 2 is indeterminate.

19. How many conditions are required in order to describe a circle? *Ans.* Three; as, for instance, the position of the centre (which depends on two conditions) and the length of the radius (compare Post. III.).

20. How is a proposition proved indirectly? *Ans.* By proving that its contradictory is false.

21. What is meant by the obverse of a proposition?

22. What propositions in Book I. are the obverse respectively of Propositions IV., V., VI., XXVII.?

23. What proposition is an instance of the *rule of identity*?

24. What are congruent figures?

25. What other name is applied to them? *Ans.* They are said to be identically equal.

26. Mention all the instances of equality which are not congruence that occur in Book I.

27. What is the difference between the symbols denoting congruence and identity?

28. Classify the properties of triangles and parallelograms proved in Book I.

29. What proposition is the converse of Prop. XXVI., Part I.?

30. Define *adjacent, exterior, interior, alternate* angles respectively.

31. What is meant by the projection of one line on another?

32. What are meant by the medians of a triangle?

33. What is meant by the third diagonal of a quadrilateral?

34. Mention some propositions in Book I. which are particular cases of more general ones that follow.

35. What is the sum of all the exterior angles of any rectilineal figure equal to?

36. How many conditions must be given in order to construct a triangle? *Ans.* Three; such as the three sides, or two sides and an angle, &c.

Exercises on Book I.

1. Any triangle is equal to the fourth part of that which is formed by drawing through each vertex a line parallel to its opposite side.

2. The three perpendiculars of the first triangle in question 1 are the perpendiculars at the middle points of the sides of the second triangle.

3. Through a given point draw a line so that the portion intercepted by the legs of a given angle may be bisected in the point.

4. The three medians of a triangle are concurrent.

5. The medians of a triangle divide each other in the ratio of 2 : 1.

6. Construct a triangle, being given two sides and the median of the third side.

7. In every triangle the sum of the medians is less than the perimeter, and greater than three-fourths of the perimeter.

8. Construct a triangle, being given a side and the two medians of the remaining sides.

9. Construct a triangle, being given the three medians.

10. The angle included between the perpendicular from the vertical angle of a triangle on the base, and the bisector of the vertical angle, is equal to half the difference of the base angles.

11. Find in two parallels two points which shall be equidistant from a given point, and whose line of connexion shall be parallel to a given line.

12. Construct a parallelogram, being given two diagonals and a side.

13. The smallest median of a triangle corresponds to the greatest side.

14. Find in two parallels two points subtending a right angle at a given point and equally distant from it.

15. The sum of the distances of any point in the base of an isosceles triangle from the equal sides is equal to the distance of either extremity of the base from the opposite side.

16. The three perpendiculars at the middle points of the sides of a triangle are concurrent. Hence prove that perpendiculars from the vertices on the opposite sides are concurrent [see Ex. 2].

17. Inscribe a lozenge in a triangle having for an angle one angle of the triangle.

18. Inscribe a square in a triangle having its base on a side of the triangle.

19. Find the locus of a point, the sum or the difference of whose distance from two fixed lines is equal to a given length.

20. The sum of the perpendiculars from any point in the interior of an equilateral triangle is equal to the perpendicular from any vertex on the opposite side.

21. The distance of the foot of the perpendicular from either extremity of the base of a triangle on the bisector of the vertical angle, from the middle point of the base, is equal to half the difference of the sides.

22. In the same case, if the bisector of the external vertical angle be taken, the distance will be equal to half the sum of the sides.

23. Find a point in one of the sides of a triangle such that the sum of the intercepts made by the other sides, on parallels drawn from the same point to these sides, may be equal to a given length.

24. If two angles have their legs respectively parallel, their bisectors are either parallel or perpendicular.

25. If lines be drawn from the extremities of the base of a triangle to the feet of perpendiculars let fall from the same points on either bisector of the vertical angle, these lines meet on the other bisector of the vertical angle.

26. The perpendiculars of a triangle are the bisectors of the angles of the triangle whose vertices are the feet of these perpendiculars.

27. Inscribe in a given triangle a parallelogram whose diagonals shall intersect in a given point.

28. Construct a quadrilateral, the four sides being given in magnitude, and the middle points of two opposite sides being given in position.

29. The bases of two or more triangles having a common vertex are given, both in magnitude and position, and the sum of the areas is given; prove that the locus of the vertex is a right line.

30. If the sum of the perpendiculars let fall from a given point on the sides of a given rectilineal figure be given, the locus of the point is a right line.

31. ABC is an isosceles triangle whose equal sides are AB, AC; $B'C'$ is any secant cutting the equal sides in B', C', so that $AB' + AC' = AB + AC$: prove that $B'C'$ is greater than BC.

32. A, B are two given points, and P is a point in a given line L; prove that the difference of AP and PB is a maximum when L bisects the angle APB; and that their sum is a minimum if it bisects the supplement.

33. Bisect a quadrilateral by a right line drawn from one of its angular points.

34. AD and BC are two parallel lines cut obliquely by AB, and perpendicularly by AC; and between these lines we draw BED, cutting AC in E, such that $ED = 2AB$; prove that the angle DBC is one-third of ABC.

35. If O be the point of concurrence of the bisectors of the angles of the triangle ABC, and if AO produced meet BC in D, and from O, OE be drawn perpendicular to BC; prove that the angle BOD is equal to the angle COE.

36. If the exterior angles of a triangle be bisected, the three external triangles formed on the sides of the original triangle are equiangular.

37. The angle made by the bisectors of two consecutive angles of a convex quadrilateral is equal to half the sum of the remaining angles; and the angle made by the bisectors of two opposite angles is equal to half the difference of the two other angles.

38. If in the construction of the figure, Proposition XLVII., EF, KG be joined,

$$EF^2 + KG^2 = 5AB^2.$$

39. Given the middle points of the sides of a convex polygon of an odd number of sides, construct the polygon.

40. Trisect a quadrilateral by lines drawn from one of its angles.

41. Given the base of a triangle in magnitude and position and the sum of the sides; prove that the perpendicular at either extremity of the base to the adjacent side, and the external bisector of the vertical angle, meet on a given line perpendicular to the base.

42. The bisectors of the angles of a convex quadrilateral form a quadrilateral whose opposite angles are supplemental. If the first quadrilateral be a parallelogram, the second is a rectangle; if the first be a rectangle, the second is a square.

43. The middle points of the sides AB, BC, CA of a triangle are respectively D, E, F; DG is drawn parallel to BF to meet EF; prove that the sides of the triangle DCG are respectively equal to the three medians of the triangle ABC.

44. Find the path of a billiard ball started from a given point which, after being reflected from the four sides of the table, will pass through another given point.

45. If two lines bisecting two angles of a triangle and terminated by the opposite sides be equal, the triangle is isosceles.

46. State and prove the Proposition corresponding to Exercise 41, when the base and difference of the sides are given.

47. If a square be inscribed in a triangle, the rectangle under its side and the sum of the base and altitude is equal to twice the area of the triangle.

48. If AB, AC be equal sides of an isosceles triangle, and if BD be a perpendicular on AC; prove that $BC^2 = 2AC \cdot CD$.

49. The sum of the equilateral triangles described on the legs of a right-angled triangle is equal to the equilateral triangle described on the hypotenuse.

50. Given the base of a triangle, the difference of the base angles, and the sum or difference of the sides; construct it.

51. Given the base of a triangle, the median that bisects the base, and the area; construct it.

52. If the diagonals AC, BD of a quadrilateral $ABCD$ intersect in E, and be bisected in the points F, G, then

$$4 \triangle EFG = (AEB + ECD) - (AED + EBC).$$

53. If squares be described on the sides of any triangle, the lines of connexion of the adjacent corners are respectively—(1) the doubles of the medians of the triangle; (2) perpendicular to them.

BOOK II.

THEORY OF RECTANGLES

Every Proposition in the Second Book has either a square or a rectangle in its enunciation. Before commencing it the student should read the following preliminary explanations: by their assistance it will be seen that this Book, which is usually considered difficult, will be rendered not only easy, but almost intuitively evident.

1. As the linear unit is that by which we express all linear measures, so the square unit is that to which all superficial measures are referred. Again, as there are different linear units in use, such as in this country, inches, feet, yards, miles, &c., and in France, metres, and their multiples or sub-multiples, so different square units are employed.

2. A square unit is the square described on a line whose length is the linear unit. Thus a square inch is the square described on a line whose length is an inch; a square foot is the square described on a line whose length is a foot, &c.

3. If we take a linear foot, describe a square on it, divide two adjacent sides each into twelve equal parts, and draw parallels to the sides, we evidently divide the square foot into square inches; and as there will manifestly be 12 rectangular parallelograms, each containing 12 square inches, the square foot contains 144 square inches.

In the same manner it can be shown that a square yard contains 9 square feet; and so in general the square described on any line contains n^2 times the square described on the nth part of the line. Thus, as a simple case, the square on a line is four times the square on its half. On account of this property the second power of a quantity is called its square; and, conversely, the square on a line AB is expressed symbolically by AB^2.

4. If a rectangular parallelogram be such that two adjacent sides contain respectively m and n linear units, by dividing one side into m and the other into n equal parts, and drawing parallels to the sides, the whole area is evidently divided into mn square units. Hence the area of the parallelogram is found by multiplying its length by its breadth, and this explains why we say (see Def. IV.) a rectangle is contained by any two adjacent sides; for if we multiply the length of one by the length of the other we have the area. Thus, if AB, AD be two adjacent sides of a rectangle, the rectangle is expressed by $AB \cdot AD$.

DEFINITIONS.

I. If a point C be taken in a line AB, the parts AC, CB are called *segments*, and C a *point of division*.

II. If C be taken in the line AB produced, AC, CB are still called the segments of the line AB; but C is called a point of *external* division.

III. A parallelogram whose angles are right angles is called a *rectangle*.

IV. A rectangle is said to be contained by any two adjacent sides. Thus the rectangle $ABCD$ is said to be contained by AB, AD, or by AB, BC, &c.

V. The rectangle contained by two separate lines such as AB and CD is the parallelogram formed by erecting a perpendicular to

AB, at A, equal to CD, and drawing parallels: the area of the rectangle will be $AB \cdot CD$.

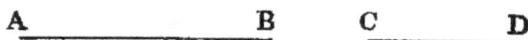

A _____ B   C _____ D

VI. In any parallelogram the figure which is composed of either of the parallelograms about a diagonal and the two complements [*see* I., XLIII.] is called a gnomon. Thus, if we take away either of the parallelograms AO, OC from the parallelogram AC, the remainder is called a *gnomon*.

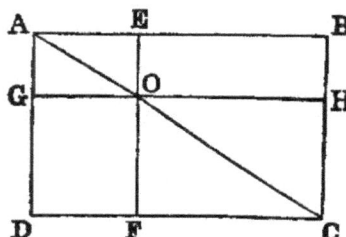

PROP. I.—THEOREM.

If there be two lines (A, BC), one of which is divided into any number of parts (BD, DE, EC), the rectangle contained by the two lines (A, BC), is equal to the sum of the rectangles contained by the undivided line (A) and the several parts of the divided line.

Dem.—Erect BF at right angles to BC [I., XI.] and make it equal to A. Complete the parallelogram BK (Def. V.). Through D, E draw DG, EH parallel to BF. Because the angles at B, D, E are right angles, each of the quadrilaterals BG, DH, EK is a rectangle. Again, since A is equal to BF (const.), the rectangle contained by A and BC is the rectangle contained by BF and BC (Def. V.); but BK is the rectangle contained by BF and BC. Hence the rectangle contained by A and BC is BK. In like manner the rectangle contained by A and BD is BG. Again, since A is equal to BF (const.), and BF is equal to DG [I. XXXIV.], A is equal to DG. Hence the rectangle contained by A and DE is the figure DH (Def. V.). In like manner the rectangle contained by A and EC is the figure EK. Hence we have the following identities:—

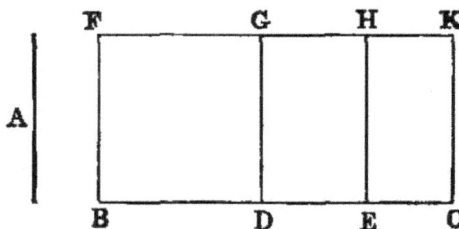

Rectangle contained by A and $BD \equiv BG$.
,,   ,, A ,, $DE \equiv DH$.
,,   ,, A ,, $EC \equiv EK$.
,,   ,, A ,, $BC \equiv BK$.

But BK is equal to the sum of BG, DH, EK (I., Axiom IX.). *Therefore the rectangle contained by A and BC is equal to the sum of the rectangles contained by A and BD, A and DE, A and EC.*

If we denote the lines BD, DE, EC by a, b, c, the Proposition asserts that the rectangle contained by A, and $a + b + c$ is equal to the sum of the rectangles contained by A and a, A

51

and b, A and c, or, as it may be written, $A(a + b + c) = Aa + Ab + Ac$. This corresponds to the distributive law in multiplication, and shows that rectangles in Geometry, and products in Arithmetic and Algebra, are subject to the same rules.

Illustration.—Suppose A to be 6 inches; BD, 5 inches; DE, 4 inches; EC, 3 inches; then BC will be 12 inches; and the rectangles will have the following values:—

$$\text{Rectangle } A \cdot BC = 6 \times 12 = 72 \text{ square inches.}$$
$$,, \qquad A \cdot BD = 6 \times \;\; 5 = 30 \qquad ,,$$
$$,, \qquad A \cdot DE = 6 \times \;\; 4 = 24 \qquad ,,$$
$$,, \qquad A \cdot EC = 6 \times \;\; 3 = 18 \qquad ,,$$

Now the sum of the three last rectangles, viz. 30, 24, 18, is 72. Hence the rectangle $A \cdot BC = A \cdot BD + A \cdot DE + A \cdot EC$.

The Second Book is occupied with the relations between the segments of a line divided in various ways. All these can be proved in the most simple manner by Algebraic Multiplication. We recommend the student to make himself acquainted with the proofs by this method as well as with those of Euclid. He will thus better understand the meaning of each Proposition.

Cor. 1.—The rectangle contained by a line and the difference of two others is equal to the difference of the rectangles contained by the line and each of the others.

Cor. 2.—The area of a triangle is equal to half the rectangle contained by its base and perpendicular.

Dem.—From the vertex C let fall the perpendicular CD. Draw EF parallel to AB, and AE, BF each parallel to CD. Then AF is the rectangle contained by AB and BF; but BF is equal to CD. Hence $AF = AB \cdot CD$; but [I. XLI.] the triangle ABC is = half the parallelogram AF. *Therefore the triangle ABC is $= \frac{1}{2}AB \cdot CD$.*

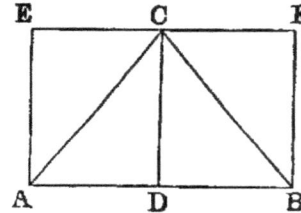

PROP. II.—THEOREM.

If a line (AB) be divided into any two parts (at C), the square on the whole line is equal to the sum of the rectangles contained by the whole and each of the segments (AC, CB).

Dem.—On AB describe the square $ABDF$ [I. XLVI.], and through C draw CE parallel to AF [I. XXXI.]. Now, since AB is equal to AF, the rectangle contained by AB and AC is equal to the rectangle contained by AF and AC; but AE is the rectangle contained by AF and AC. *Hence the rectangle contained by AB and AC is equal to AE. In like manner the rectangle contained by AB and CB is equal to the figure CD. Therefore the sum of the two rectangles $AB \cdot AC$, $AB \cdot CB$ is equal to the square on AB.*

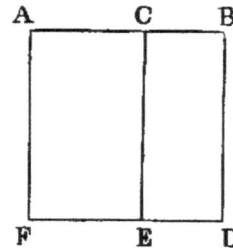

Or thus: $\qquad\qquad AB = AC + CB,$

and $\qquad\qquad\qquad\qquad AB = AB.$

Hence, multiplying, we get $\qquad AB^2 = AB \cdot AC + AB \cdot CB.$

52

This Proposition is the particular case of I. when the divided and undivided lines are equal, hence it does not require a separate Demonstration.

PROP. III.—THEOREM.

If a line (AB) be divided into two segments (at C), the rectangle contained by the whole line and either segment (CB) is equal to the square on that segment together with the rectangle contained by the segments.

Dem.—On BC describe the square $BCDE$ [I. XLVI.]. Through A draw AF parallel to CD: produce ED to meet AF in F. Now since CB is equal to CD, the rectangle contained by AC, CB is equal to the rectangle contained by AC, CD; but the rectangle contained by AC, CD is the figure AD. Hence the rectangle $AC.CB$ is equal to the figure AD, and the square on CB is the figure CE. Hence the rectangle $AC.CB$, together with the square on CB, is equal to the figure AE.

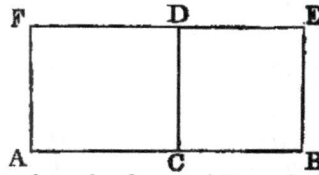

Again, since CB is equal to BE, the rectangle $AB.CB$ is equal to the rectangle $AB.BE$; but the rectangle $AB.BE$ is equal to the figure AE. Hence the rectangle $AB.CB$ is equal to the figure AE. And since things which are equal to the same are equal to one another, *the rectangle AC.CB, together with the square on CB, is equal to the rectangle AB.CB.*

Or thus:
$$AB = AC + CB,$$
$$CB = CB.$$

Hence
$$AB.CB = AC.CB + CB^2.$$

Prop. III. is the particular case of Prop. I., when the undivided line is equal to a segment of the divided line.

PROP. IV.—THEOREM.

If a line (AB) be divided into any two parts (at C), the square on the whole line is equal to the sum of the squares on the parts (AC, CB), together with twice their rectangle.

Dem.—On AB describe a square $ABDE$. Join EB; through C draw CF parallel to AE, intersecting BE in G; and through G draw HI parallel to AB.

Now since AE is equal to AB, the angle ABE is equal to AEB [I. V.]; but since BE intersects the parallels AE, CF, the angle AEB is equal to CGB [I. XXIX.]. Hence the angle CBG is equal to CGB, and therefore [I. VI.] CG is equal to CB; but CG is equal to BI and CB to GI. Hence the figure $CBIG$ is a lozenge, and the angle CBI is right. Hence (I., Def. XXX.) it is a square. In like manner the figure $EFGH$ is a square.

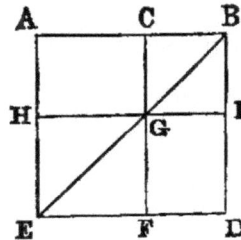

53

Again, since CB is equal to CG, the rectangle $AC.CB$ is equal to the rectangle $AC.CG$; but $AC.CG$ is the figure AG (Def. IV.). Therefore the rectangle $AC.CB$ is equal to the figure AG. Now the figures AG, GD are equal [I. XLIII.], being the complements about the diagonal of the parallelogram AD. Hence the parallelograms AG, GD are together equal to twice the rectangle $AC.CB$. Again, the figure HF is the square on HG, and HG is equal to AC. Therefore HF is equal to the square on AC, and CI is the square on CB; but the whole figure AD, which is the square on AB, is the sum of the four figures HF, CI, AG, GD. *Therefore the square on AB is equal to the sum of the squares on AC, CB, and twice the rectangle $AC.CB$.*

Or thus: On AB describe the square $ABDE$, and cut off AH, EG, DF each equal to CB. Join CF, FG, GH, HC. Now the four $\triangle$s ACH, CBF, FDG, GEH are evidently equal; therefore their sum is equal to four times the $\triangle ACH$; but the $\triangle ACH$ is half the rectangle $AC.AH$ (I. *Cor.* 2)—that is, equal to half the rectangle $AC.CB$. Therefore the sum of the four triangles is equal to $2AC.CB$.

Again, the figure $CFGH$ is a square [I. XLVI., *Cor.* 3], and equal to $AC^2 + AH^2$ [I. XLVII.]—that is, equal to $AC^2 + CB^2$. *Hence the whole figure $ABDE = AC^2 + CB^2 + 2AC.CB$.*

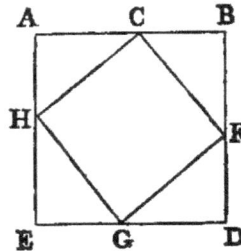

Or thus:
$$AB = AC + CB.$$
Squaring, we get
$$AB^2 = AC^2 + 2AC.CB + CB^2.$$

Cor. 1.—The parallelograms about the diagonal of a square are squares.

Cor. 2.—The square on a line is equal to four times the square on its half.

For let $AB = 2AC$, then $AB^2 = 4AC^2$.

This *Cor.* may be proved by the First Book thus: Erect CD at right angles to AB, and make $CD = AC$ or CB. Join AD, DB.

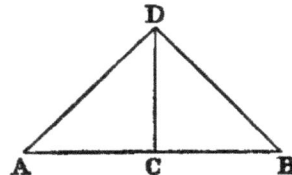

Then $$AD^2 = AC^2 + CD^2 = 2AC^2$$

In like manner, $$DB^2 = 2CB^2;$$

therefore $$AD^2 + DB^2 = 2AC^2 + 2CB^2 = 4AC^2.$$

But since the angle ADB is right, $AD^2 + DB^2 = AB^2$;

therefore $$AB^2 = 4AC^2.$$

Cor. 3.—If a line be divided into any number of parts, the square on the whole is equal to the sum of the squares on all the parts, together with twice the sum of the rectangles contained by the several distinct pairs of parts.

Exercises.

1. Prove Proposition IV. by using Propositions II. and III.

2. If from the vertical angle of a right-angled triangle a perpendicular be let fall on the hypotenuse, its square is equal to the rectangle contained by the segments of the hypotenuse.

3. From the hypotenuse of a right-angled triangle portions are cut off equal to the adjacent sides; prove that the square on the middle segment is equal to twice the rectangle contained by the extreme segments.

4. In any right-angled triangle the square on the sum of the hypotenuse and perpendicular, from the right angle on the hypotenuse, exceeds the square on the sum of the sides by the square on the perpendicular.

5. The square on the perimeter of a right-angled triangle is equal to twice the rectangle contained by the sum of the hypotenuse and one side, and the sum of the hypotenuse and the other side.

PROP. V.—THEOREM.

If a line (AB) be divided into two equal parts (at C), and also into two unequal parts (at D), the rectangle ($AD.DB$) contained by the unequal parts, together with the square on the part (CD) between the points of section, is equal to the square on half the line.

Dem.—On CB describe the square $CBEF$ [I. XLVI.]. Join BF. Through D draw DG parallel to CF, meeting BF in H. Through H draw KM parallel to AB, and through A draw AK parallel to CL [I. XXXI.].

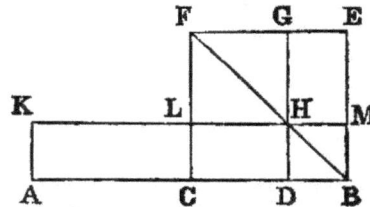

The parallelogram CM is equal to DE [I. XLIII., *Cor.* 2]; but AL is equal to CM [I. XXXVI.], because they are on equal bases AC, CB, and between the same parallels; therefore AL is equal to DE: to each add CH, and we get the parallelogram AH equal to the gnomon CMG; but AH is equal to the rectangle $AD.DH$, and therefore equal to the rectangle $AD.DB$, since DH is equal to DB [IV., *Cor.* 1]; therefore the rectangle $AD.DB$ is equal to the gnomon CMG, and the square on CD is equal to the figure LG. *Hence the rectangle $AD.DB$, together with the square on CD, is equal to the whole figure $CBEF$—that is, to the square on CB.*

Or thus:
$$AD = AC + CD = BC + CD;$$
$$DB = \qquad\qquad BC - CD;$$
therefore
$$AD.BD = (BC + CD)(BC - CD) = BC^2 - CD^2.$$
Hence
$$AD.BD + CD^2 = BC^2.$$

Cor. 1.—The rectangle $AD.DB$ is the rectangle contained by the sum of the lines AC, CD and their difference; and we have proved it equal to the difference between the square on AC and the square on CD. *Hence the difference of the*

55

squares on two lines is equal to the rectangle contained by their sum and their difference.

Cor. 2.—The perimeter of the rectangle AH is equal to $2AB$, and is therefore independent of the position of the point D on the line AB; and the area of the same rectangle is less than the square on half the line by the square on the segment between D and the middle point of the line; therefore, when D is the middle point, the rectangle will have the maximum area. *Hence, of all rectangles having the same perimeter, the square has the greatest area.*

Exercises.

1. Divide a given line so that the rectangle contained by its parts may have a maximum area.

2. Divide a given line so that the rectangle contained by its segments may be equal to a given square, not exceeding the square on half the given line.

3. The rectangle contained by the sum and the difference of two sides of a triangle is equal to the rectangle contained by the base and the difference of the segments of the base, made by the perpendicular from the vertex.

4. The difference of the sides of a triangle is less than the difference of the segments of the base, made by the perpendicular from the vertex.

5. The difference between the square on one of the equal sides of an isosceles triangle, and the square on any line drawn from the vertex to a point in the base, is equal to the rectangle contained by the segments of the base.

6. The square on either side of a right-angled triangle is equal to the rectangle contained by the sum and the difference of the hypotenuse and the other side.

PROP. VI.—Theorem.

If a line (AB) be bisected (at C), and divided externally in any point (D), the rectangle $(AD \cdot BD)$ contained by the segments made by the external point, together with the square on half the line, is equal to the square on the segment between the middle point and the point of external division.

Dem.—On CD describe the square $CDFE$ [I. xlvi.], and join DE; through B draw BHG parallel to CE [I. xxxi.], meeting DE in H; through H draw KLM parallel to AD; and through A draw AK parallel to CL. Then because AC is equal to CB, the rectangle AL is equal to CH [I. xxxvi.]; but the complements CH, HF are equal [I. xliii.]; therefore AL is equal to HF. To each of these equals add CM and LG, and we get AM and LG equal to the square $CDFE$; but AM is equal to the rectangle $AD \cdot DM$, and therefore equal to the rectangle $AD \cdot DB$, since DB is equal to DM; also LG is equal to the square on CB, and $CDFE$ is the square on CD. *Hence the rectangle $AD \cdot DB$, together with the square on CB, is equal to the square on CD.*

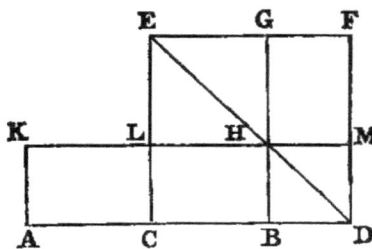

Or thus:—

Dem.—On CB describe the square $CBEF$ [I. XLVI.]. Join BF. Through D draw DG parallel to CF, meeting FB produced in H. Through H draw KM parallel to AB. Through A draw AK parallel to CL [I. XXXI.].

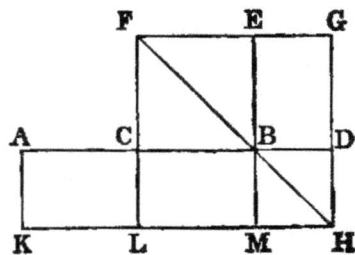

The parallelogram CM is equal to DE [I. XLIII.]; but AL is equal to CM [I. XXXVI.], because they are on equal bases AC, CB, and between the same parallels; therefore AL is equal to DE. To each add CH, and we get the parallelogram AH equal to the gnomon CMG; but AH is equal to the rectangle $AD.DH$, and therefore equal to the rectangle $AD.DB$, since DH is equal to DB [IV., *Cor.* 1]; therefore the rectangle $AD.DB$ is equal to the gnomon CMG, and the square on CB is the figure CE. *Therefore the rectangle $AD.DB$, together with the square on CB, is equal to the whole figure $LHGF$—that is, equal to the square on LH or to the square on CD.*

Or thus:

$$AD = AC + CD = CD + CB;$$
$$BD = \qquad\qquad CD - CB.$$

Hence
$$AD.DB = (CD + CB)(CD - CB) = CD^2 - CB^2;$$

therefore
$$AD.DB + CB^2 = CD^2.$$

Exercises.

1. Show that Proposition VI. is reduced to Proposition V. by producing the line in the opposite direction.

2. Divide a given line externally, so that the rectangle contained by its segments may be equal to the square on a given line.

3. Given the difference of two lines and the rectangle contained by them; find the lines.

4. The rectangle contained by any two lines is equal to the square on half the sum, minus the square on half the difference.

5. Given the sum or the difference of two lines and the difference of their squares; find the lines.

6. If from the vertex C of an isosceles triangle a line CD be drawn to any point in the base produced, prove that $CD^2 - CB^2 = AD.DB$.

7. Give a common enunciation which will include Propositions V. and VI.

PROP. VII.—THEOREM.

If a right line (AB) be divided into any two parts (at C), the sum of the squares on the whole line (AB) and either segment (CB) is equal to twice the rectangle $(2AB.CB)$ contained by the whole line and that segment, together with the square on the other segment.

Dem.—On AB describe the square $ABDE$. Join BE. Through C draw CG parallel to AE, intersecting BE in F. Through F draw HK parallel to AB.

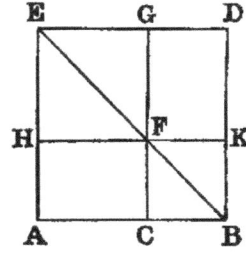

Now the square AD is equal to the three figures AK, FD, and GH: to each add the square CK, and we have the sum of the squares AD, CK equal to the sum of the three figures AK, CD, GH; but CD is equal to AK; therefore the sum of the squares AD, CK is equal to twice the figure AK, together with the figure GH. Now AK is the rectangle $AB.BK$; but BK is equal to BC; therefore AK is equal to the rectangle $AB.BC$, and AD is the square on AB; CK the square on CB; and GH is the square on HF, and therefore equal to the square on AC. Hence the sum of the squares on AB and BC is equal to twice the rectangle $AB.BC$, together with the square on AC.

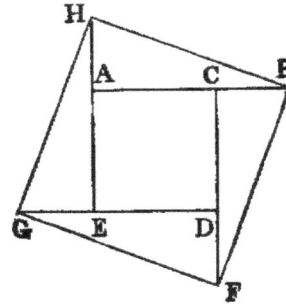

Or thus: On AC describe the square $ACDE$. Produce the sides CD, DE, EA, and make each produced part equal to CB. Join BF, FG, GH, HB. Then the figure $BFGH$ is a square [I. XLVI., Ex. 3], and it is equal to the square on AC, together with the four equal triangles HAB, BCF, FDG, GEH. Now [I. XLVII.], the figure $BFGH$ is equal to the sum of the squares on AB, AH—that is, equal to the sum of the squares on AB, BC; and the sum of the four triangles is equal to twice the rectangle $AB.BC$, for each triangle is equal to half the rectangle $AB.BC$. Hence *the sum of the squares on AB, BC is equal to twice the rectangle $AB.BC$, together with the square on AC.*

Or thus: $$AC = AB - BC;$$
therefore $$AC^2 = AB^2 - 2AB.BC + BC^2;$$
therefore $$AC^2 + 2AB.BC = AB^2 + BC^2.$$

Comparison of IV. and VII.
By IV., square on sum = sum of squares + twice rectangle.
By VII., square on difference = sum of squares−twice rectangle.

Cors. from IV. and VII.

1. Square on the sum, the sum of the squares, and the square on the difference of any two lines, are in arithmetical progression.

2. Square on the sum + square on the difference of any two lines = twice the sum of the squares on the lines (Props. IX. and X.).

3. The square on the sum − the square on the difference of any two lines = four times the rectangle under lines (Prop. VIII.).

PROP. VIII.–THEOREM.

If a line (AB) be divided into two parts (at C), the square on the sum of the whole line (AB) and either segment (BC) is equal to four times the rectangle contained by the whole line (AB) and that segment, together with the square on the other segment (AC).

58

Dem.—Produce AB to D. Make BD equal to BC. On AD describe the square $AEFD$ [I. XLVI.]. Join DE. Through C, B draw CH, BL parallel to AE [I. XXXI.], and through K, I draw MN, PO parallel to AD.

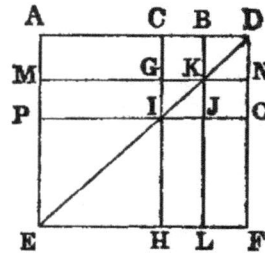

Since CO is the square on CD, and CK the square on CB, and CB is the half of CD, CO is equal to four times CK [IV., Cor. 1]. Again, since CG, GI are the sides of equal squares, they are equal [I. XLVI., Cor. 1]. Hence the parallelogram AG is equal to MI [I. XXXVI.]. In like manner IL is equal to JF; but MI is equal to IL [I. XLIII.]. Therefore the four figures AG, MI, IL, JF are all equal; hence their sum is equal to four times AG; and the square CO has been proved to be equal to four times CK. Hence the gnomon AOH is equal to four times the rectangle AK—that is, equal to four times the rectangle $AB.BC$, since BC is equal to BK.

Again, the figure PH is the square on PI, and therefore equal to the square on AC. Hence the whole figure AF, that is, *the square on AD, is equal to four times the rectangle $AB.BC$, together with the square on AC.*

Or thus: Produce BA to D, and make $AD = BC$. On DB describe the square $DBEF$. Cut off BG, EI, FL each equal to BC. Through A and I draw lines parallel to DF, and through G and L, lines parallel to AB.

Now it is evident that the four rectangles. AG, GI, IL, LA are all equal; but AG is the rectangle $AB.BG$ or $AB.BC$. Therefore the sum of the four rectangles is equal to $4AB.BC$. Again, the figure NP is evidently equal to the square on AC. Hence the whole figure, which is the square on BD, or *the square on the sum of AB and BC, is equal to $4AB.BC + AC^2$.*

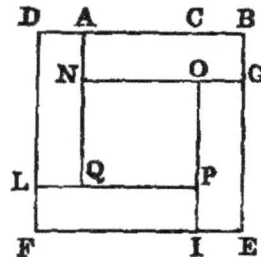

Or thus:
$$AB + BC = AC + 2BC;$$
therefore
$$(AB + BC)^2 = AC^2 + 4AC.CB + 4BC^2$$
$$= AC^2 + 4(AC + CB).CB$$
$$= AC^2 + 4AB.BC.$$

Direct sequence from V. or VI.

Since by V. or VI. the rectangle contained by any two lines is = the square on half their sum − the square on half their difference; therefore four times the rectangle contained by any two lines = the square on their sum − the square on their difference.

Direct sequence of VIII. from IV. and VII.

By IV., the square on the sum = the sum of the squares + twice the rectangle.

By VII., the square on the difference = the sum of the squares − twice the rectangle. Therefore, by subtraction, the square on the sum − the square on the difference = four times the rectangle.

Exercises.

1. In the figure [I. XLVII.] if EF, GK be joined, prove $EF^2 - CO^2 = (AB + BO)^2$.
2. Prove $GK^2 - EF^2 = 3AB(AO - BO)$.
3.[1] Given the difference of two lines $= R$, and their rectangle $= 4R^2$; find the lines.

PROP. IX.—THEOREM.

If a line (AB) be bisected (at C) and divided into two unequal parts (at D), the sum of the squares on the unequal parts (AD, DB) is double the sum of the squares on half the line (AC), and on the segment (CD) between the points of section.

Dem.—Erect CE at right angles to AB, and make it equal to AC or CB. Join AE, EB. Draw DF parallel to CE, and FG parallel to CD. Join AF.

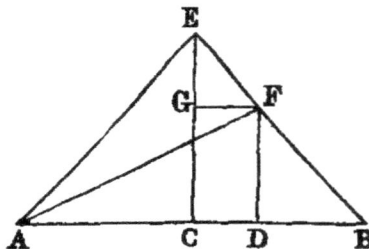

Because AC is equal to CE, and the angle ACE is right, the angle CEA is half a right angle. In like manner the angles CEB, CBE are half right angles; therefore the whole angle AEF is right. Again, because GF is parallel to CB, and CE intersects them, the angle EGF is equal to ECB; but ECB is right (const.); therefore EGF is right; and GEF has been proved to be half a right angle; therefore the angle GFE is half a right angle [I. XXXII.]. Therefore [I. VI.] GE is equal to GF. In like manner FD is equal to DB.

Again, since AC is equal to CE, AC^2 is equal to CE^2; but AE^2 is equal to $AC^2 + CE^2$ [I. XLVII.]. Therefore AE^2 is equal to $2AC^2$. In like manner EF^2 is equal to $2GF^2$ or $2CD^2$. Therefore $AE^2 + EF^2$ is equal to $2AC^2 + 2CD^2$; but $AE^2 + EF^2$ is equal to AF^2 [I. XLVII.]. Therefore AF^2 is equal to $2AC^2 + 2CD^2$.

Again, since DF is equal to DB, DF^2 is equal to DB^2: to each add AD^2, and we get $AD^2 + DF^2$ equal to $AD^2 + DB^2$; but $AD^2 + DF^2$ is equal to AF^2; therefore AF^2 is equal to $AD^2 + DB^2$; and we have proved AF^2 equal to $2AC^2 + 2CD^2$. *Therefore $AD^2 + DB^2$ is equal to $2AC^2 + 2CD^2$.*

Or thus: $AD = AC + CD$; $DB = AC - CD$.

Square and add, and we get $AD^2 + DB^2 = 2AC^2 + 2CD^2$.

---

[1]Ex. 3 occurs in the solution of the problem of the inscription of a regular polygon of seventeen sides in a circle. *See* note C.

Exercises.

1. The sum of the squares on the segments of a line of given length is a minimum when it is bisected.

2. Divide a given line internally, so that the sum of the squares on the parts may be equal to a given square, and state the limitation to its possibility.

3. If a line AB be bisected in C and divided unequally in D,

$$AD^2 + DB^2 = 2AD.DB + 4CD^2.$$

4. Twice the square on the line joining any point in the hypotenuse of a right-angled isosceles triangle to the vertex is equal to the sum of the squares on the segments of the hypotenuse.

5. If a line be divided into any number of parts, the continued product of all the parts is a maximum, and the sum of their squares is a minimum when all the parts are equal.

PROP. X.—THEOREM.

If a line (AB) be bisected (at C) and divided externally (at D), the sum of the squares on the segments (AD, DB) made by the external point is equal to twice the square on half the line, and twice the square on the segment between the points of section.

Dem.—Erect CE at right angles to AB, and make it equal to AC or CB. Join AE, EB. Draw DF parallel to CE, and produce EB. Now since DF is parallel to EC, the angle BDF is = to BCE [I. XXIX.], and [I. XV.] the angle DBF is = to EBC; but the sum of the angles BCE, EBC is less than two right angles [I. XVII.]; therefore the sum of the angles BDF, DBF is less than two right angles, and therefore [I., Axiom XII.] the lines EB, DF, if produced, will meet. Let them meet in F. Through F draw FG parallel to AB, and produce EC to meet it in G. Join AF.

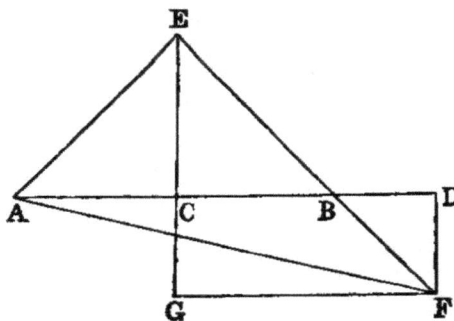

Because AC is equal to CE, and the angle ACE is right, the angle CEA is half a right angle. In like manner the angles CEB, CBE are half right angles; therefore the whole angle AEF is right. Again, because GF is parallel to CB, and GE intersects them, the angle EGF is equal to ECB [I. XXIX.]; but ECB is right (const.); therefore EGF is right, and GEF has been proved to be half a right angle; therefore [I. XXXII.] GFE is half a right angle, and therefore [I. VI.] GE is equal to GF. In like manner FD is equal to DB.

Again, since AC is equal to CE, AC^2 is equal to CE^2; but AE^2 is equal to $AC^2 + CE^2$ [I. XLVII.]; therefore AE^2 is equal to $2AC^2$. In like manner EF^2 is equal to $2GF^2$ or $2CD^2$; therefore $AE^2 + EF^2$ is equal to $2AC^2 + 2CD^2$; but $AE^2 + EF^2$ is equal to AF^2 [I. XLVII.]. Therefore AF^2 is equal to $2AC^2 + 2CD^2$.

Again, since DF is equal to DB, DF^2 is equal to DB^2: to each add AD^2, and we get $AD^2 + DF^2$ equal to $AD^2 + DB^2$; but $AD^2 + DF^2$ is equal to

61

AF^2; therefore AF^2 is equal to $AD^2 + DB^2$; and AF^2 has been proved equal to $2AC^2 + 2CD^2$. *Therefore $AD^2 + DB^2$ is equal to $2AC^2 + 2CD^2$.*

Or thus:
$$AD = CD + AC,$$
$$BD = CD - AC.$$

Square and add, and we get $AD^2 + BD^2 = 2CD^2 + 2AC^2$.

The following enunciations include Propositions IX. and X.:—

1. *The square on the sum of any two lines plus the square on their difference equal twice the sum of their squares.*

2. *The sum of the squares on any two lines it equal to twice the square on half the sum plus twice the square on half the difference of the lines.*

3. *If a line be cut into two unequal parts, and also into two equal parts, the sum of the squares on the two unequal parts exceeds the sum of the squares on the two equal parts by the sum of the squares of the two differences between the equal and unequal parts.*

Exercises

1. Given the sum or the difference of any two lines, and the sum of their squares; find the lines.

2. The sum of the squares on two sides AC, CB of a triangle is equal to twice the square on half the base AB, and twice the square on the median which bisects AB.

3. If the base of a triangle be given both in magnitude and position, and the sum of the squares on the sides in magnitude, the locus of the vertex is a circle.

4. If in the $\triangle ABC$ a point D in the base BC be such that

$$BA^2 + BD^2 = CA^2 + CD^2;$$

prove that the middle point of AD is equally distant from B and C.

5. The sum of the squares on the sides of a parallelogram is equal to the sum of the squares on its diagonals.

PROP. XI.—PROBLEM.

To divide a given finite line (AB) into two segments (in H), so that the rectangle $(AB . BH)$ contained by the whole line and one segment may be equal to the square on the other segment.

Sol.—On AB describe the square $ABDC$ [I. XLVI.]. Bisect AC in E. Join BE. Produce EA to F, and make EF equal to EB. On AF describe the square $AFGH$. *H is the point required.*

Dem.—Produce GH to K. Then because CA is bisected in E, and divided externally in F, the rectangle $CF . AF$, together with the square on EA, is equal to the square on EF [VI.]; but EF is equal to EB (const.);

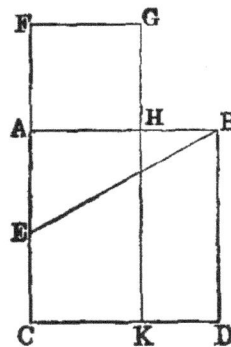

therefore the rectangle $CF \cdot AF$, together with EA^2, is equal to EB^2—that is [I. XLVII.] equal to $EA^2 + AB^2$. Rejecting EA^2, which is common, we get the rectangle $CF \cdot AF$ equal to AB^2. Again, since AF is equal to FG, being the sides of a square, the rectangle $CF \cdot AF$ is equal to $CF \cdot FG$—that is, to the figure CG; and AB^2 is equal to the figure AD; therefore CG is equal to AD. Reject the part AK, which is common, and we get the figure FH equal to the figure HD; but HD is equal to the rectangle $AB \cdot BH$, because BD is equal to AB, and FH is the square on AH. *Therefore the rectangle $AB \cdot BH$ is equal to the square on AH.*

DEF.—A line divided as in this Proposition is said to be divided in "*extreme and mean ratio.*"

Cor. 1.—The line CF is divided in "extreme and mean ratio" at A.

Cor. 2.—If from the greater segment CA of CF we take a segment equal to AF, it is evident that CA will be divided into parts respectively equal to AH, HB. Hence, if a line be divided in extreme and mean ratio, the greater segment will be cut in the same manner by taking on it a part equal to the less; and the less will be similarly divided by taking on it a part equal to the difference, and so on, &c.

Cor. 3.—Let AB be divided in "extreme and mean ratio" in C, then it is evident (*Cor.* 2) that AC is greater than CB. Cut off $CD =$

CB; then (*Cor.* 2) AC is cut in "extreme and mean ratio" at D, and CD is greater than AD. Next, cut off DE equal to AD, and in the same manner we have DE greater than EC, and so on. Now since CD is greater than AD, it is evident that CD is not a common measure of AC and CB, and therefore not a common measure of AB and AC. In like manner AD is not a common measure of AC and CD, and therefore not a common measure of AB and AC. Hence, no matter how far we proceed we cannot arrive at any remainder which will be a common measure of AB and AC. Hence, *the parts of a line divided in "extreme and mean ratio" are incommensurable.*

Exercises.

1. Cut a line externally in "extreme and mean ratio."

2. The difference between the squares on the segments of a line divided in "extreme and mean ratio" is equal to their rectangle.

3. In a right-angled triangle, if the square on one side be equal to the rectangle contained by the hypotenuse and the other side, the hypotenuse is cut in "extreme and mean ratio" by the perpendicular on it from the right angle.

4. If AB be cut in "extreme and mean ratio" at C, prove that

(1) $AB^2 + BC^2 = 3AC^2$.

(2) $(AB + BC)^2 = 5AC^2$.

5. The three lines joining the pairs of points G, B; F, D; A, K, in the construction of Proposition XI., are parallel.

6. If CH intersect BE in O, AO is perpendicular to CH.

7. If CH be produced, it meets BF at right angles.

8. ABC is a right-angled triangle having $AB = 2AC$: if AH be made equal to the difference between BC and AC, AB is divided in "extreme and mean ratio" at H.

PROP. XII.—Theorem.

In an obtuse-angled triangle (ABC), the square on the side (AB) subtending the obtuse angle exceeds the sum of the squares on the sides (BC, CA) containing the obtuse angle, by twice the rectangle contained by either of them (BC), and its continuation (CD) to meet a perpendicular (AD) on it from the opposite angle.

Dem.—Because BD is divided into two parts in C, we have

$$BD^2 = BC^2 + CD^2 + 2BC \cdot CD \text{ [iv.]}$$

and

$$AD^2 = AD^2.$$

Hence, adding, since [I. xlvii.] $BD^2 + AD^2 = AB^2$, and $CD^2 + AD^2 = CA^2$, we get

$$AB^2 = BC^2 + CA^2 + 2BC \cdot CD.$$

Therefore AB^2 is greater than $BC^2 + CA^2$ by $2BC \cdot CD$.

The foregoing proof differs from Euclid's only in the use of symbols. I have found by experience that pupils more readily understand it than any other method.

Or thus: By the First Book: Describe squares on the three sides. Draw AE, BF, CG perpendicular to the sides of the squares. Then it can be proved exactly as in the demonstration of [I. xlvii.], that the rectangle BG is equal to BE, AG to AF, and CE to CF. Hence the sum of the two squares on AC, CB is less than the square on AB by twice the rectangle CE; that is, by twice the rectangle $BC \cdot CD$.

Cor. 1.—If perpendiculars from A and B to the opposite sides meet them in H and D, the rectangle $AC \cdot CH$ is equal to the rectangle $BC \cdot CD$.

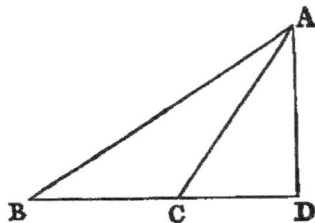

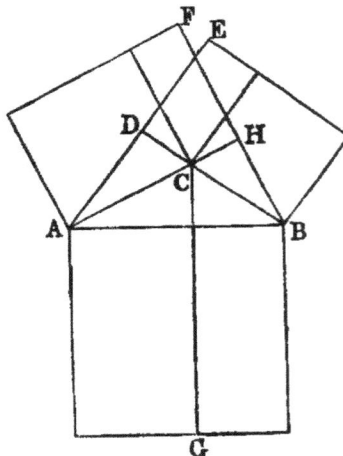

Exercises.

1. If the angle ACB of a triangle be equal to twice the angle of an equilateral triangle, $AB^2 = BC^2 + CA^2 + BC \cdot CA$.

2. $ABCD$ is a quadrilateral whose opposite angles B and D are right, and AD, BC produced meet in E; prove $AE \cdot DE = BE \cdot CE$.

3. ABC is a right-angled triangle, and BD is a perpendicular on the hypotenuse AC; Prove $AB \cdot DC = BD \cdot BC$.

4. If a line AB be divided in C so that $AC^2 = 2CB^2$; prove that $AB^2 + BC^2 = 2AB \cdot AC$.

5. If AB be the diameter of a semicircle, find a point C in AB such that, joining C to a fixed point D in the circumference, and erecting a perpendicular CE meeting the circumference in E, $CE^2 - CD^2$ may be equal to a given square.

6. If the square of a line CD, drawn from the angle C of an equilateral triangle ABC to a point D in the side AB produced, be equal to $2AB^2$; prove that AD is cut in "extreme and mean ratio" at B.

64

PROP. XIII.—THEOREM.

In any triangle (ABC), the square on any side subtending an acute angle (C) is less than the sum of the squares on the sides containing that angle, by twice the rectangle (BC, CD) contained by either of them (BC) and the intercept (CD) between the acute angle and the foot of the perpendicular on it from the opposite angle.

Dem.—Because BC is divided into two segments in D,

$$BC^2 + CD^2 = BD^2 + 2BC \cdot CD \text{ [VII.]};$$

and

$$AD^2 = AD^2.$$

Hence, adding, since

$$CD^2 + AD^2 = AC^2 \text{ [I. XLVII.]},$$

and

$$BD^2 + AD^2 = AB^2,$$

we get

$$BC^2 + AC^2 = AB^2 + 2BC \cdot CD.$$

Therefore AB^2 is less than $BC^2 + AC^2$ by $2BC \cdot CD$.

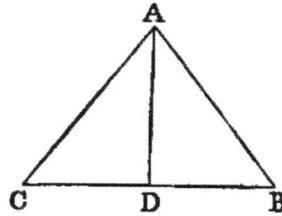

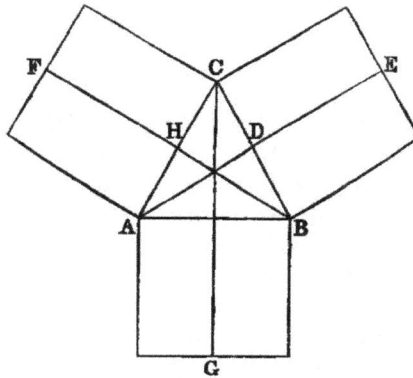

Or thus: Describe squares on the sides. Draw AE, BF, CG perpendicular to the sides; then, as in the demonstration of [I. XLVII.], the rectangle BG is equal to BE; AG to AF, and CE to CF. Hence the sum of the squares on AC, CB exceeds the square on AB by twice CE—that is, by $2BC \cdot CD$.

Observation.—By comparing the proofs of the pairs of Props. IV. and VII.; V. and VI.; IX. and X.; XII. and XIII., it will be seen that they are virtually identical. In order to render this identity more apparent, we have made some slight alterations in the usual proofs. The pairs of Propositions thus grouped are considered in Modern Geometry not as distinct, but each pair is regarded as one Proposition.

1. If the angle C of the $\triangle\ ACB$ be equal to an angle of an equilateral $\triangle$, $AB^2 = AC^2 + BC^2 - AC\,.\,BC$.

2. The sum of the squares on the diagonals of a quadrilateral, together with four times the square on the line joining their middle points, is equal to the sum of the squares on its sides.

3. Find a point C in a given line AB produced, so that $AC^2 + BC^2 = 2AC\,.\,BC$.

PROP. XIV.—PROBLEM.

To construct a square equal to a given rectilineal figure (X).

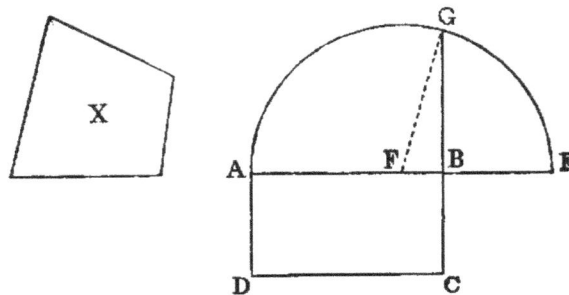

Sol.—Construct [I. XLV.] the rectangle AC equal to X. Then, if the adjacent sides AB, BC be equal, AC is a square, and the problem is solved; if not, produce AB to E, and make BE equal to BC; bisect AE in F; with F as centre and FE as radius, describe the semicircle AGE; produce CB to meet it in G. *The square described on BG will be equal to X.*

Dem.—Join FG. Then because AE is divided equally in F and unequally in B, the rectangle $AB\,.\,BE$, together with FB^2 is equal to FE^2 [V.], that is, to FG^2; but FG^2 is equal to $FB^2 + BG^2$ [I. XLVII.]. Therefore the rectangle $AB\,.\,BE + FB^2$ is equal to $FB^2 + BG^2$. Reject FB^2, which is common, and we have the rectangle $AB\,.\,BE = BG^2$; but since BE is equal to BC, the rectangle $AB\,.\,BE$ is equal to the figure AC. Therefore BG^2 is equal to the figure AC, *and therefore equal to the given rectilineal figure (X).*

Cor.—The square on the perpendicular from any point in a semicircle on the diameter is equal to the rectangle contained by the segments of the diameter.

1. Given the difference of the squares on two lines and their rectangle; find the lines.

2. Divide a given line, so that the rectangle contained by another given line and one segment may be equal to the square on the other segment.

Questions for Examination on Book II.

1. What is the subject-matter of Book II.? *Ans.* Theory of rectangles.

2. What is a rectangle? A gnomon?

3. What is a square inch? A square foot? A square perch? A square mile? *Ans.* The square described on a line whose length is an inch, a foot, a perch, &c.

4. What is the difference between linear and superficial measurement? *Ans.* Linear measurement has but one dimension; superficial has two.

5. When is a line said to be divided internally? When externally?

6. How is the area of a rectangle found?

7. How is a line divided so that the rectangle contained by its segments may be a maximum?

8. How is the area of a parallelogram found?

9. What is the altitude of a parallelogram whose base is 65 metres and area 1430 square metres?

10. How is a line divided when the sum of the squares on its segments is a minimum?

11. The area of a rectangle is 108.60 square metres and its perimeter is 48.20 linear metres; find its dimensions.

12. What Proposition in Book II. expresses the distributive law of multiplication?

13. On what proposition is the rule for extracting the square root founded?

14. Compare I. XLVII. and II. XII. and XIII.

15. If the sides of a triangle be expressed by x^2+1, x^2-1, and $2x$ linear units, respectively; prove that it is right-angled.

16. How would you construct a square whose area would be exactly an acre? Give a solution by I. XLVII.

17. What is meant by incommensurable lines? Give an example from Book II.

18. Prove that a side and the diagonal of a square are incommensurable.

19. The diagonals of a lozenge are 16 and 30 metres respectively; find the length of a side.

20. The diagonal of a rectangle is 4.25 perches, and its area is 7.50 square perches; what are its dimensions?

21. The three sides of a triangle are 8, 11, 15; prove that it has an obtuse angle.

22. The sides of a triangle are 13, 14, 15; find the lengths of its medians; also the lengths of its perpendiculars, and prove that all its angles are acute.

23. If the sides of a triangle be expressed by m^2+n^2, m^2-n^2, and $2mn$ linear units, respectively; prove that it is right-angled.

24. If on each side of a square containing 5.29 square perches we measure from the corners respectively a distance of 1.5 linear perches; find the area of the square formed by joining the points thus found.

Exercises on Book II.

1. The squares on the diagonals of a quadrilateral are together double the sum of the squares on the lines joining the middle points of opposite sides.

2. If the medians of a triangle intersect in O, $AB^2+BC^2+CA^2 = 3(OA^2+OB^2+OC^2)$.

3. Through a given point O draw three lines OA, OB, OC of given lengths, such that their extremities may be collinear, and that $AB = BC$.

4. If in any quadrilateral two opposite sides be bisected, the sum of the squares on the other two sides, together with the sum of the squares on the diagonals, is equal to the sum of the squares on the bisected sides, together with four times the square on the line joining the points of bisection.

5. If squares be described on the sides of any triangle, the sum of the squares on the lines joining the adjacent corners is equal to three times the sum of the squares on the sides of the triangle.

6. Divide a given line into two parts, so that the rectangle contained by the whole and one segment may be equal to any multiple of the square on the other segment.

7. If P be any point in the diameter AB of a semicircle, and CD any parallel chord, then

$$CP^2 + PD^2 = AP^2 + PB^2.$$

8. If A, B, C, D be four collinear points taken in order,

$$AB \cdot CD + BC \cdot AD = AC \cdot BD.$$

9. Three times the sum of the squares on the sides of any pentagon exceeds the sum of the squares on its diagonals, by four times the sum of the squares on the lines joining the middle points of the diagonals.

10. In any triangle, three times the sum of the squares on the sides is equal to four times the sum of the squares on the medians.

11. If perpendiculars be drawn from the angular points of a square to any line, the sum of the squares on the perpendiculars from one pair of opposite angles exceeds twice the rectangle of the perpendiculars from the other pair by the area of the square.

12. If the base AB of a triangle be divided in D, so that $mAD = nBD$, then

$$mAC^2 + nBC^2 = mAD^2 + nDB^2 + (m+n)CD^2.$$

13. If the point D be taken in AB produced, so that $mAD = nDB$, then

$$mAC^2 - nBC^2 = mAD^2 - nDB^2 + (m-n)CD^2.$$

14. Given the base of a triangle in magnitude and position, and the sum or the difference of m times the square on one side and n times the square on the other side, in magnitude, the locus of the vertex is a circle.

15. Any rectangle is equal to half the rectangle contained by the diagonals of squares described on its adjacent sides.

16. If A, B, C. &c., be any number of fixed points, and P a variable point, find the locus of P, if $AP^2 + BP^2 + CP^2 +$ &c., be given in magnitude.

17. If the area of a rectangle be given, its perimeter is a minimum when it is a square.

18. If a transversal cut in the points A, C, B three lines issuing from a point D, prove that

$$BC \cdot AD^2 + AC \cdot BD^2 - AB \cdot CD^2 = AB \cdot BC \cdot CA.$$

19. Upon the segments AC, CB of a line AB equilateral triangles are described: prove that if D, D' be the centres of circles described about these triangles, $6DD'^2 = AB^2 + AC^2 + CB^2$.

20. If a, b, p denote the sides of a right-angled triangle about the right angle, and the perpendicular from the right angle on the hypotenuse, $\dfrac{1}{a^2} + \dfrac{1}{b^2} = \dfrac{1}{p^2}$.

21. If, upon the greater segment AB of a line AC, divided in extreme and mean ratio, an equilateral triangle ABD be described, and CD joined, $CD^2 = 2AB^2$.

22. If a variable line, whose extremities rest on the circumferences of two given concentric circles, subtend a right angle at any fixed point, the locus of its middle point is a circle.

BOOK III.

THEORY OF THE CIRCLE

DEFINITIONS.

I. Equal circles are those whose radii are equal.

This is a theorem, and not a definition. For if two circles have equal radii, they are evidently congruent figures, and therefore equal. From this way of proving this theorem Props. XXVI.–XXIX. follow as immediate inferences.

II. A chord of a circle is the line joining two points in its circumference.

If the chord be produced both ways, the whole line is called a secant, and each of the parts into which a secant divides the circumference is called an arc—the greater the *major conjugate arc*, and the lesser the *minor conjugate arc*.—NEWCOMB.

III. A right line is said to touch a circle when it meets the circle, and, being produced both ways, does not cut it; the line is called a *tangent* to the circle, and the point where it touches it the *point of contact.*

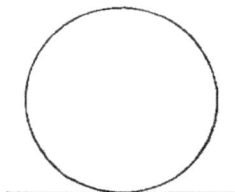

In Modern Geometry a curve is considered as made up of an infinite number of points, which are placed in order along the curve, and then the secant through two consecutive points is a tangent. Euclid's definition for a tangent is quite inadequate for any curve but the circle, and those derived from it by projection (the conic sections); and even for these the modern definition is better.

IV. Circles are said to touch one another when they meet, but do not intersect. There are two species of contact:—
1. When each circle is external to the other.
2. When one is inside the other.

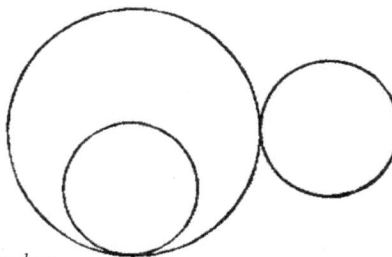

The following is the modern definition of curve-contact:— *When two curves have two, three, four, &c., consecutive points common, they have contact of the first, second, third, &c., orders.*

V. A segment of a circle is a figure bounded by a chord and one of the arcs into which it divides the circumference.

VI. Chords are said to be equally distant from the centre when the perpendiculars drawn to them from the centre are equal.

VII. The angle contained by two lines, drawn from any point in the circumference of a segment to the extremities of its chord, is called an *angle in the segment.*

69

VIII. The angle of a segment is the angle contained between its chord and the tangent at either extremity.

A theorem is tacitly assumed in this Definition, namely, that the angles which the chord makes with the tangent at its extremities are equal. We shall prove this further on.

IX. An angle in a segment is said to *stand* on its conjugate arc.

X. Similar segments of circles are those that contain equal angles.

XI. A sector of a circle is formed of two radii and the arc included between them.

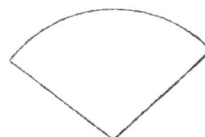

To a pair of radii may belong either of the two conjugate arcs into which their ends divide the circle.—NEWCOMB.

XII. Concentric circles are those that have the same centre.

XIII. Points which lie on the circumference of a circle are said to be *concyclic*.

XIV. A *cyclic quadrilateral* is one which is inscribed in a circle.

XV. It will be proper to give here an explanation of the extended meaning of the word *angle* in Modern Geometry. This extension is necessary in Trigonometry, in Mechanics—in fact, in every application of Geometry, and has been partly given in I. Def. IX.

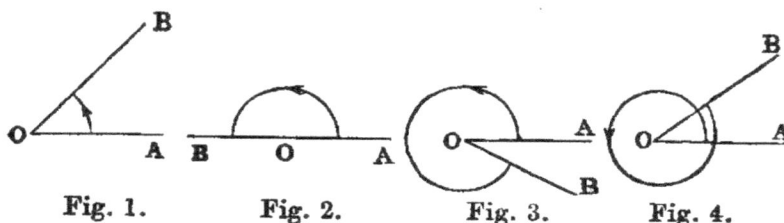

Fig. 1.   Fig. 2.   Fig. 3.   Fig. 4.

Thus, if a line OA revolve about the point O, as in figures 1, 2, 3, 4, until it comes into the position OB, the amount of the *rotation* from OA to OB is called an *angle*. From the diagrams we see that in fig. 1 it is less than two right angles; in fig. 2 it is equal to two right angles; in fig. 3 greater than two right angles, but less than four; and in fig. 4 it is greater than four right angles. The arrow-heads denote the direction or *sense*, as it is technically termed, in which the line OA turns. It is usual to call the direction indicated in the above figures *positive*, and the opposite *negative*. A line such as OA, which turns about a fixed point, is called a *ray*, and then we have the following definition:—

XVI. A ray which turns in the sense *opposite* to the hands of a watch describes a *positive angle*, and one which turns in the *same* direction as the hands, a *negative angle*.

PROP. I.—PROBLEM.

To find the centre of a given circle (ADB).

Sol.—Take any two points A, B in the circumference. Join AB. Bisect it in C. Erect CD at right angles to AB. Produce DC to meet the circle again in E. Bisect DE in F. *Then F is the centre.*

Dem.—If possible, let any other point G be the centre. Join GA, GC, GB. Then in the triangles ACG, BCG we have AC equal to CB (const.), CG common, and the base GA equal to GB, because they are drawn from G, which is, by hypothesis, the centre, to the circumference. Hence [I. VIII.] the angle ACG is equal to the adjacent angle BCG, and therefore [I. Def. XIII.] each is a right angle; but the angle ACD is right (const.); therefore ACD is equal to ACG—a part equal to the whole—which is absurd. Hence no point can be the centre which is not in the line DE. *Therefore F, the middle point of DE, must be the centre.*

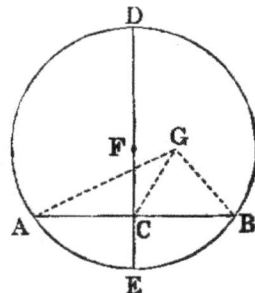

The foregoing proof may be abridged as follows:—
Because ED bisects AB at right angles, every point equally distant from, the points A, B must lie in ED [I. x. Ex. 2]; but the centre is equally distant from A and B; hence the centre must be in ED; *and since it must be equally distant from E and D, it must be the middle point of DE.*

Cor. 1.—The line which bisects any chord of a circle perpendicularly passes through the centre of the circle.

Cor. 2.—The locus of the centres of the circles which pass through two fixed points is the line bisecting at right angles that connecting the two points.

Cor. 3.—If A, B, C be three points in the circumference of a circle, the lines bisecting perpendicularly the chords AB, BC intersect in the centre.

PROP. II.—THEOREM.

If any two points (A, B) be taken in the circumference of a circle—1. The segment (AB) of the indefinite line through these points which lies between them falls within the circle. 2. The remaining parts of the line are without the circle.

Dem.—1. Let C be the centre. Take any point D in AB. Join CA, CD, CB. Now the angle ADC is [I. XVI.] greater than ABC; but the angle ABC is equal to CAB [I. V.], because the triangle CAB is isosceles; therefore the angle ADC is greater than CAD. Hence AC is greater than CD [I. XIX.]; therefore CD is less than the radius of the circle, consequently the point D must be within the circle (note on I. Def. XXIII.).

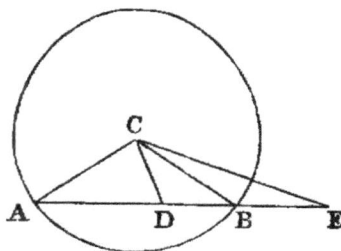

In the same manner *every other point between A and B lies within the circle.*

2. Take any point E in AB produced either way. Join CE. Then the angle ABC is greater than AEC [I. XVI.]; therefore CAB is greater than AEC. Hence CE is greater than CA, *and the point E is without the circle.*

We have added the second part of this Proposition. The indirect proof given of the first part in several editions of Euclid is very inelegant; it is one of those absurd things which give many students a dislike to Geometry.

Cor. 1.—Three collinear points cannot be concyclic.

Cor. 2.—A line cannot meet a circle in more than two points.

Cor. 3.—The circumference of a circle is everywhere concave towards the centre.

PROP. III.–Theorem.

If a line (AB) passing through the centre of a circle bisect a chord (CD), which does not pass through the centre, it cuts it at right angles. 2. If it cuts it at right angles, it bisects it.

Dem.—1. Let O be the centre of the circle. Join OC, OD. Then the triangles CEO, DEO have CE equal to ED (hyp.), EO common, and OC equal to OD, because they are radii of the circle; hence [I. VIII.] the angle CEO is equal to DEO, and they are adjacent angles. Therefore [I. Def. XIII.] each is a right angle. *Hence AB cuts CD at right angles.*

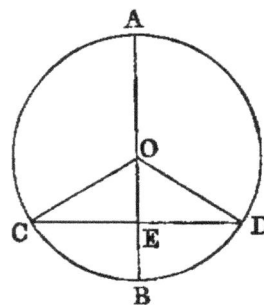

2. The same construction being made: because OC is equal to OD, the angle OCD is equal to ODC [I. V.], and CEO is equal to DEO (hyp.), because each is right. Therefore the triangles CEO, DEO have two angles in one respectively equal to two angles in the other, and the side EO common. Hence [I. XXVI.] the side CE is equal to ED. *Therefore CD is bisected in E.*

2. May be proved as follows:—

$$OC^2 = OE^2 + EC^2 \text{ [I. XLVII.]}, \text{ and } OD^2 = OE^2 + ED^2;$$

but
$$OC^2 = OD^2; \therefore OE^2 + EC^2 = OE^2 + ED^2.$$

Hence
$$EC^2 = ED^2, \text{ and } EC = ED.$$

Observation.—The three theorems, namely, *Cor.* 1., *Prop.* I., and Parts 1, 2, of Prop. III., are so related, that any one being proved directly, the other two follow by the Rule of Identity.

Cor. 1.—The line which bisects perpendicularly one of two parallel chords of a circle bisects the other perpendicularly.

Cor. 2.—The locus of the middle points of a system of parallel chords of a circle is the diameter of the circle perpendicular to them all.

Cor. 3.—If a line intersect two concentric circles, its intercepts between the circles are equal.

Cor. 4.—The line joining the centres of two intersecting circles bisects their common chord perpendicularly.

1. If a chord of a circle subtend a right angle at a given point, the locus of its middle point is a circle.

2. Every circle passing through a given point, and having its centre on a given line, passes through another given point.

3. Draw a chord in a given circle which shall subtend a right angle at a given point, and be parallel to a given line.

PROP. IV.—THEOREM.

Two chords of a circle (AB, CD) which are not both diameters cannot bisect each other, though either may bisect the other.

Dem.—Let O be the centre. Let AB, CD intersect in E; then since AB, CD are not both diameters, join OE. If possible let AE be equal to EB, and CE equal to ED. Now, since OE passing through the centre bisects AB, which does not pass through the centre, it is at right angles to it; therefore the angle AEO is right. In like manner the angle CEO is right. Hence AEO is equal to CEO—that is, a part equal to the whole—which is absurd. *Therefore AB and CD do not bisect each other.*

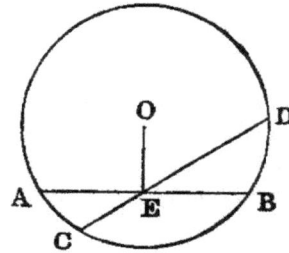

Cor.—If two chords of a circle bisect each other, they are both diameters.

PROP. V.—THEOREM.

If two circles (ABC, ABD) cut one another in any point (A), they are not concentric.

Dem.—If possible let them have a common centre at O. Join OA, and draw any other line OD, cutting the circles in C and D respectively. Then because O is the centre of the circle ABC, OA is equal to OC. Again, because O is the centre of the circle ABD, OA is equal to OD. Hence OC is equal to OD—a part equal to the whole—which is absurd. *Therefore the circles are not concentric.*

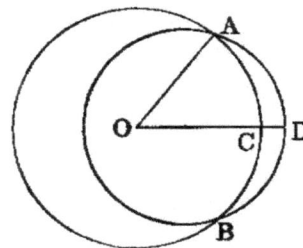

1. If two non-concentric circles intersect in one point, they must intersect in another point. For, let O, O' be the centres, A the point of intersection; from A let fall the $\perp AC$ on the line OO'. Produce AC to B, making $BC = CA$: then B is another point of intersection.

2. Two circles cannot have three points in common without wholly coinciding.

PROP. VI.—THEOREM.

If one circle (ABC) touch another circle (ADE) internally in any point (A), it is not concentric with it.

Dem.—If possible let the circles be concentric, and let O be the centre of each. Join OA, and draw any other line OD, cutting the circles in the points B, D respectively. Then because O is the centre of each circle (hyp.), OB and OD are each equal to OA; therefore OB is equal to OD, which is impossible. *Hence the circles cannot have the same centre.*

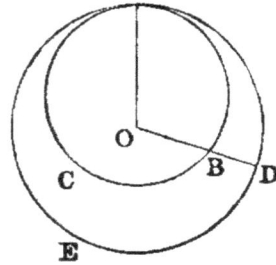

PROP. VII.—THEOREM.

If from any point (P) within a circle, which is not the centre, lines (PA, PB, PC, &c.), one of which passes through the centre, be drawn to the circumference, then—1. The greatest is the line (PA) which passes through the centre. 2. The production (PE) of this in the opposite direction is the least. 3. Of the others, that which is nearest to the line through the centre is greater than every one more remote. 4. Any two lines making equal angles with the diameter on opposite sides are equal. 5. More than two equal right lines cannot be drawn from the given point (P) to the circumference.

Dem.—1. Let O be the centre. Join OB. Now since O is the centre, OA is equal to OB: to each add OP, and we have AP equal to the sum of OB, OP; but the sum of OB, OP is greater than PB [I. xx.]. *Therefore PA is greater than PB.*

2. Join OD. Then [I. xx.] the sum of OP, PD is greater than OD; but OD is equal to OE [I. Def. xxx.]. Therefore the sum of OP, PD is greater than OE. Reject OP, which is common, and we have *PD greater than PE.*

3. Join OC; then two triangles POB, POC have the side OB equal to OC [I. Def. xxx.], and OP common; but the angle POB is greater than POC; therefore [I. xxiv.] the base PB is greater than PC. In like manner PC is *greater than PD.*

4. Make at the centre O the angle POF equal to POD. Join PF. Then the triangles POD, POF have the two sides OP, OD in one respectively equal to the sides OP, OF in the other, and the angle POD equal to the angle POF; hence PD is equal to PF [I. iv.], and the angle OPD equal to the angle OPF. *Therefore PD and PF make equal angles with the diameter.*

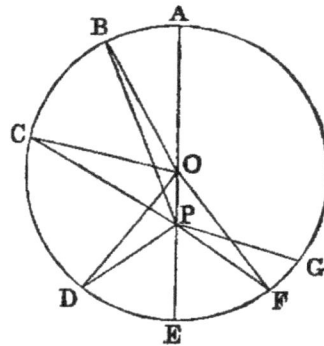

74

5. A third line cannot be drawn from P equal to either of the equal lines PD, PF. If possible let PG be equal to PD, then PG is equal to PF—that is, the line which is nearest to the one through the centre is equal to the one which is more remote, which is impossible. *Hence three equal lines cannot be drawn from P to the circumference.*

Cor. 1.—If two equal lines PD, PF be drawn from a point P to the circumference of a circle, the diameter through P bisects the angle DPF formed by these lines.

Cor. 2.—If P be the common centre of circles whose radii are PA, PB, PC, &c., then—1. The circle whose radius is the maximum line (PA) lies outside the circle ADE, and touches it in A [Def. IV.]. 2. The circle whose radius is the minimum line (PE) lies inside the circle ADE, and touches it in E. 3. A circle having any of the remaining lines (PD) as radius cuts ADE in two points (D, F).

Observation.—Proposition VII. affords a good illustration of the following important definition (see *Sequel to Euclid,* p. 13):—If a geometrical magnitude varies its position continuously according to any law, and if it retains the same value throughout, it is said to be a constant, such as the radius of a circle revolving round the centre; but if it goes on increasing for some time, and then begins to decrease, it is said to be *a maximum* at the end of the increase. Thus, in the foregoing figure, PA, supposed to revolve round P and meet the circle, is a maximum. Again, if it decreases for some time, and then begins to increase, it is *a minimum* at the commencement of the increase. Thus PE, supposed as before to revolve round P and meet the circle, is a minimum. Proposition VIII. will give other illustrations.

PROP. VIII.—THEOREM.

If from any point (P) outside a circle, lines (PA, PB, PC, &c.) be drawn to the concave circumference, then— 1. The maximum is that which passes through the centre. 2. Of the others, that which is nearer to the one through the centre is greater than the one more remote. Again, if lines be drawn to the convex circumference—3. The minimum is that whose production passes through the centre. 4. Of the others, that which is nearer to the minimum is less than one more remote. 5. From the given point (P) there can be drawn two equal lines to the concave or the convex circumference, both of which make equal angles with the line passing through the centre. 6. More than two equal lines cannot be drawn from the given point (P) to either circumference.

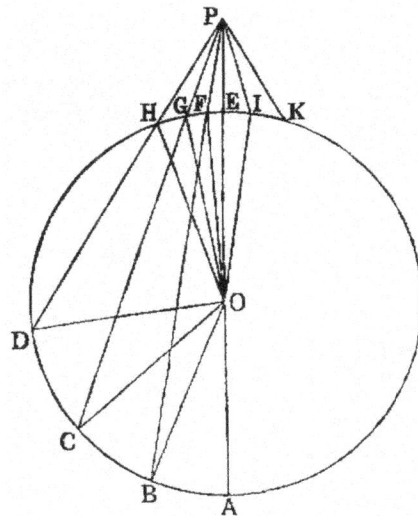

Dem.—1. Let O be the centre. Join OB. Now since O is the centre, OA is equal to OB: to each add OP, and we have AP equal to the sum of OB, OP;

but the sum of OB, OP is greater than BP [I. XX.]. *Therefore AP is greater than BP.*

2. Join OC, OD. The two triangles BOP, COP have the side OB equal to OC, and OP common, and the angle BOP greater than COP; *therefore the base BP is greater than CP* [I. XXIV.]. *In like manner CP is greater than DP,* &c.

3. Join OF. Now in the triangle OFP the sum of the sides OF, FP is greater than OP [I. XX.]; but OF is equal to OE [I. Def. XXX.]. Reject them, and FP *will remain greater than EP*.

4. Join OG, OH. The two triangles GOP, FOP have two sides GO, OP in one respectively equal to two sides FO, OP in the other; but the angle GOP is greater than FOP; *therefore* [I. XXIV.] *the base GP is greater than FP. In like manner HP is greater than GP.*

5. Make the angle POI equal POF [I. XXIII.]. Join IP. Now the triangles IOP, FOP have two sides IO, OP in one respectively equal to two sides FO, OP in the other, and the angle IOP equal to FOP (const.); *therefore* [I. IV.] *IP is equal to FP.*

6. *A third line cannot be drawn from P equal to either of the lines IP, FP.* For if possible let PK be equal to PF; then PK is equal to PI—that is, one which is nearer to the minimum equal to one more remote—*which is impossible.*

Cor. 1.—If PI be produced to meet the circle again in L, PL is equal to PB.

Cor. 2.—If two equal lines be drawn from P to either the convex or concave circumference, the diameter through P bisects the angle between them, and the parts of them intercepted by the circle are equal.

Cor. 3.—If P be the common centre of circles whose radii are lines drawn from P to the circumference of HDE, then—1. The circle whose radius is the minimum line (PE) has contact of the *first kind* with ADE [Def. IV.]. 2. The circle whose radius is the maximum line (PA) has contact of the *second kind*. 3. A circle having any of the remaining lines (PF) as radius intersects HDE in two points (F, I).

PROP. IX.—THEOREM.

A point (P) within a circle (ABC), from which more than two equal lines (PA, PB, PC, &c.) can be drawn to the circumference, is the centre.

Dem.—If P be not the centre, let O be the centre. Join OP, and produce it to meet the circle in D and E; then DE is the diameter, and P is a point in it which is not the centre: therefore [VII.] only two equal lines can be drawn from P to the circumference; but three equal lines are drawn (hyp.), which is absurd. *Hence P must be the centre.*

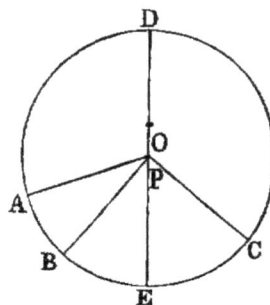

Or thus: Since the lines AP, BP are equal, the line bisecting the angle APB [VII. *Cor.* 1] must pass through the centre: in like manner the line bisecting the angle BPC must pass through the centre. *Hence the point of intersection of these bisectors, that is, the point P, must be the centre.*

PROP. X.—THEOREM.

If two circles have more than two points common, they must coincide.

Dem.—Let X be one of the circles; and if possible let another circle Y have three points, A, B, C, in common with X, without coinciding with it. Find P, the centre of X. Join PA, PB, PC. Then since P is the centre of X, the three lines PA, PB, PC are equal to one another.

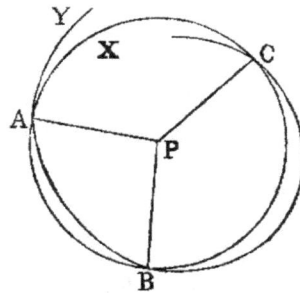

Again, since Y is a circle and P a point, from which three equal lines PA, PB, PC can be drawn to its circumference, P must be the centre of Y. *Hence X and Y are concentric, which* [V.] *is impossible.*

Cor.—Two circles not coinciding cannot have more than two points common. Compare I., Axiom X., that two right lines not coinciding cannot have more than one point common.

PROP. XI.—THEOREM.

If one circle (CPD) touch another circle (APB) internally at any point P, the line joining the centres must pass through that point.

Dem.—Let O be the centre of APB. Join OP. I say the centre of the smaller circle is in the line OP. If not, let it be in any other position such as E. Join OE, EP, and produce OE through E to meet the circles in the points C, A. Now since E is a point in the diameter of the larger circle between the centre and A, EA is less than EP [VII. 2]; but EP is equal to EC (hyp.), being radii of the smaller circle. Hence EA is less than EC; which is impossible; consequently the centre of the smaller circle must be in the line OP. Let it be H; then we see *that the line joining the centres passes through the point P.*

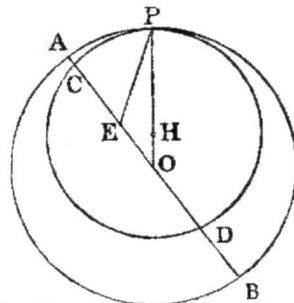

Or thus: Since EP is a line drawn from a point within the circle APB to the circumference, but not forming part of the diameter through E, the circle whose centre is E and radius EP cuts [VII., *Cor.* 2] APB in P, but it touches it (hyp.) also in P, which is impossible. *Hence the centre of the smaller circle CPD must be in the line OP.*

77

*If two circles (PCF, PDE) have external contact at any point P, the line
joining their centres must pass through that point.*

Dem.—Let *A* be the centre of one of
the circles. Join *AP*, and produce it to
meet the second circle again in *E*. I say
the centre of the second circle is in the line
PE. If not, let it be elsewhere, as at *B*.
Join *AB*, intersecting the circles in *C* and
D, and join *BP*. Now since *A* is the centre
of the circle *PCF*, *AP* is equal to *AC*; and since *B* is the centre of the circle
PDE, *BP* is equal to *BD*. Hence the sum of the lines *AP*, *BP* is equal to
the sum of the lines *AC*, *DB*; but *AB* is greater than the sum of *AC* and *DB*;
therefore *AB* is greater than the sum of *AP*, *PB*—that is, one side of a triangle
greater than the sum of the other two–which [I. xx.] is impossible. Hence the
centre of the second circle must be in the line *PE*. Let it be *G*, and we see that
the line through the centres passes through the point P.

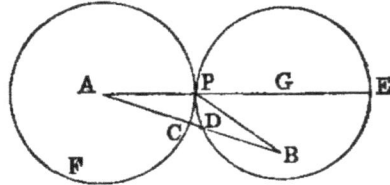

Or thus: Since *BP* is a line drawn from a point without the circle *PCF*
to its circumference, and when produced does not pass through the centre, the
circle whose centre is *B* and radius *BP* must cut the circle *PCF* in *P* [viii.,
Cor. 3]; but it touches it (hyp.) also in *P*, which is impossible. *Hence the centre
of the second circle must be in the line PE.*

Observation.—Propositions xi, xii., may both be included in one enunciation as follows:—
"If two circles touch each other at any point, the centres and that point are collinear." And this
latter Proposition is a limiting case of the theorem given in Proposition iii., *Cor.* 4, that "The
line joining the centres of two intersecting circles bisects the common chord perpendicularly."

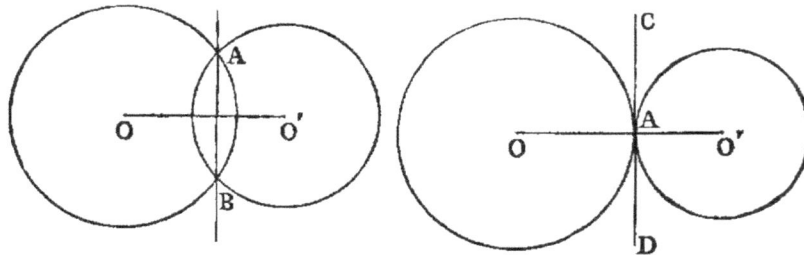

Suppose the circle whose centre is *O* and one of the points of intersection *A* to remain
fixed, while the second circle turns round that point in such a manner that the second point
of intersection *B* becomes ultimately consecutive to *A*; then, since the line *OO'* always bisects
AB, we see that when *B* ultimately becomes consecutive to *A*, the line *OO'* passes through
A. In consequence of the motion, the common chord will become in the limit a tangent to
each circle, as in the second diagram.—Comberousse, *Géométrie Plane*, page 57.

Cor. 1.—If two circles touch each other, their point of contact is the union of two points
of intersection. Hence a contact counts for two intersections.

Cor. 2.—If two circles touch each other at any point, they cannot have any other common point. For, since two circles cannot have more than two points common [X.], and that the point of contact is equivalent to two common points, circles that touch cannot have any other point common. The following is a formal proof of this Proposition:—Let O, O' be the centres of the two circles, A the point of contact, and let O' lie between O and A; take any other point B in the circumference of O. Join $O'B$; then [VII.] $O'B$ is greater than $O'A$; therefore the point C is outside the circumference of the smaller circle. Hence B cannot be common to both circles. In like manner, they cannot have any other common point but A.

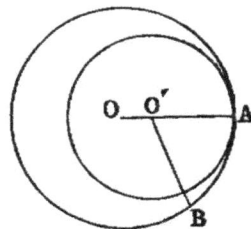

PROP. XIII.—THEOREM.

Two circles cannot have double contact, that it, cannot touch each other in two points.

Dem.—1. If possible let two circles touch each other at two points A and B. Now since the two circles touch each other in A, the line joining their centres passes through A [XI.]. In like manner, it passes through B. Hence the centres and the points A, B are in one right line; therefore AB is a diameter of each circle. Hence, if AB be bisected in E, E must be the centre of each circle—that is, the circles are concentric—*which* [V.] *is impossible.*

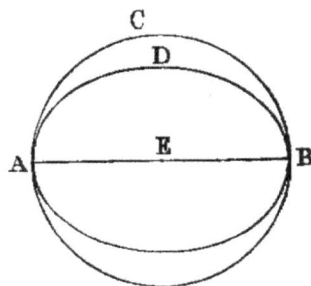

2. If two circles touched each other externally in two distinct points, then [XII.] the line joining the centres should pass through each point, *which is impossible.*

Or thus: Draw a line bisecting AB at right angles. Then this line [I., *Cor.* 1] must pass through the centre of each circle, and therefore [XI. XII.] must pass through each point of contact, which is impossible. *Hence two circles cannot have double contact.*

This Proposition is an immediate inference from the theorem [XII., *Cor.* 1], that a point of contact counts for two intersections, for then two contacts would be equivalent to four intersections; but there cannot be more than two intersections [X.]. It also follows from Prop. XII., *Cor.* 2, that if two circles touch each other in a point A, they cannot have any other point common; *hence they cannot touch again in B.*

Exercises.

1. If a variable circle touch two fixed circles externally, the difference of the distances of its centre from the centres of the fixed circles is equal to the difference or the sum of their radii, according as the contacts are of the same or of opposite species (Def. IV.).

2. If a variable circle be touched by one of two fixed circles internally, and touch the other fixed circle either externally or internally, the sum of the distances of its centre from the centres of the fixed circles is equal to the sum or the difference of their radii, according as the contact with the second circle is of the first or second kind.

3. If through the point of contact of two touching circles any secant be drawn cutting the circles again in two points, the radii drawn to these points are parallel.

79

4. If two diameters of two touching circles be parallel, the lines from the point of contact to the extremities of one diameter pass through the extremities of the other.

PROP. XIV.—Theorem.

In equal circles—1. equal chords (AB, CD) are equally distant from the centre. 2. chords which are equally distant from the centre are equal.

Dem.—1. Let O be the centre. Draw the perpendiculars OE, OF. Join AO, CO. Then because AB is a chord in a circle, and OE is drawn from the centre cutting it at right angles, it bisects it [III.]; therefore AE is the half of AB. In like manner, CF is the half of CD; but AB is equal to CD (hyp.). Therefore AE is equal to CF [I., Axiom VII.]. And because E is a right angle, AO^2 is equal to $AE^2 + EO^2$. In like manner, CO^2 is equal to $CF^2 + FO^2$; but AO^2 is equal to CO^2. Therefore $AE^2 + EO^2$ is equal to $CF^2 + FO^2$; and AE^2 has been proved equal to CF^2. Hence EO^2 is equal to FO^2; therefore EO is equal to FO. *Hence AB, CD are* (Def. VI.) *equally distant from the centre.*

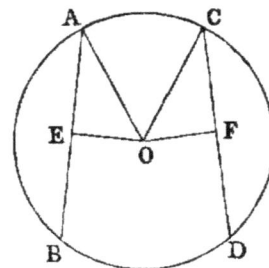

2. Let EO be equal to FO, it is required to prove AB equal to CD. The same construction being made, we have, as before, $AE^2 + EO^2$ equal to $CF^2 + FO^2$; but EO^2 is equal to FO^2 (hyp.). Hence AE^2 is equal to CF^2, and AE is equal to CF; but AB is double of AE, and CD double of CF. *Therefore AB is equal to CD.*

Exercise.

If a chord of given length slide round a fixed circle—1. the locus of its middle point is a circle; 2. the locus of any point fixed in the chord is a circle.

PROP. XV.—Theorem.

The diameter (AB) is the greatest chord in a circle; and of the others, the chord (CD) which is nearer to the centre is greater than (EF) one more remote, and the greater is nearer to the centre than the less.

Dem.—1. Join OC, OD, OE, and draw the perpendiculars OG, OH; then because O is the centre, OA is equal to OC [I., Def. XXXII.], and OB is equal to OD. Hence AB is equal to the sum of OC and OD; but the sum of OC, OD is greater than CD [I. XX.]. *Therefore AB is greater than CD.*

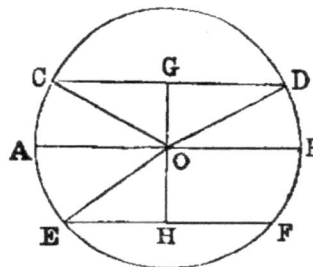

2. Because the chord CD is nearer to the centre than EF, OG is less than OH; and since the

triangles OGC, OHE are right-angled, we have $OC^2 = OG^2 + GC^2$, and $OE^2 = OH^2 + HE^2$; therefore $OG^2 + GC^2 = OH^2 + HE^2$; but OG^2 is less than OH^2; therefore GC^2 is greater than HE^2, and GC is greater than HE, but CD and EF are the doubles of GC and HE. *Hence CD is greater than EF.*

3. Let CD be greater than EF, it is required to prove that OG is less than OH.

As before, we have $OG^2 + GC^2$ equal to $OH^2 + HE^2$; but CG^2 is greater than EH^2; therefore OG^2 is less than OH^2. *Hence OG is less than OH.*

Exercises.

1. The shortest chord which can be drawn through a given point within a circle is the perpendicular to the diameter which passes through that point.

2. Through a given point, within or without a given circle, draw a chord of length equal to that of a given chord.

3. Through one of the points of intersection of two circles draw a secant—1. the sum of whose segments intercepted by the circles shall be a maximum; 2. which shall be of any length less than that of the maximum.

4. Three circles touch each other externally at A, B, C; the chords AB, AC of two of them are produced to meet the third again in the points D and E; prove that DE is a diameter of the third circle, and parallel to the line joining the centres of the others.

PROP. XVI.—THEOREM.

1. *The perpendicular (BI) to the diameter (AB) of a circle at its extremity (B) touches the circle at that point.* 2. *Any other line (BH) through the same point cuts the circle.*

Dem.—1. Take any point I, and join it to the centre C. Then because the angle CBI is a right angle, CI^2 is equal to $CB^2 + BI^2$ [I. XLVII.]; therefore CI^2 is greater than CB^2. Hence CI is greater than CB, and the point I [note on I., Def. XXXII.] is without the circle. In like manner, every other point in BI, except B, is without the circle. *Hence, since BI meets the circle at B, but does not cut it, it must touch it.*

2. To prove that BH, which is not perpendicular to AB, cuts the circle. Draw CG perpendicular to HB. Now BC^2 is equal to $CG^2 + GB^2$. Therefore BC^2 is greater than CG^2, and BC is greater than CG. Hence [note on I., Def. XXXII.] the point G must be within the circle, and consequently the line BG produced must meet the circle again, *and must therefore cut it.*

This Proposition may be proved as follows:

At every point on a circle the tangent is perpendicular to the radius.

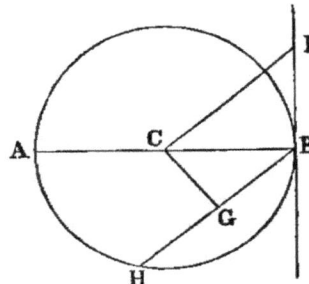

81

Let P and Q be two consecutive points on the circumference. Join CP, CQ, PQ; produce PQ both ways. Now since P and Q are consecutive points, PQ is a tangent (Def. III.). Again, the sum of the three angles of the triangle CPQ is equal to two right angles; but the angle C is infinitely small, and the others are equal. Hence each of them is a right angle. *Therefore the tangent is perpendicular to the diameter.*

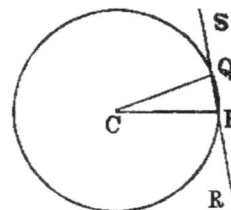

Or thus: A tangent is a limiting position of a secant, namely, when the secant moves out until the two points of intersection with the circle become consecutive; but the line through the centre which bisects the part of the secant within the circle [III.] is perpendicular to it. *Hence, in the limit the tangent is perpendicular to the line from the centre to the point of contact.*

Or again: The angle CPR is always equal to CQS; hence, when P and Q come together each is a right angle, *and the tangent is perpendicular to the radius.*

Exercises.

1. If two circles be concentric, all chords of the greater which touch the lesser are equal.
2. Draw a parallel to a given line to touch a given circle.
3. Draw a perpendicular to a given line to touch a given circle.
4. Describe a circle having its centre at a given point—1. and touching a given line; 2. and touching a given circle. How many solutions of this case?
5. Describe a circle of given radius that shall touch two given lines. How many solutions?
6. Find the locus of the centres of a system of circles touching two given lines.
7. Describe a circle of given radius that shall touch a given circle and a given line, or that shall touch two given circles.

PROP. XVII.—PROBLEM.

From a given point (P) without a given circle (BCD) to draw a tangent to the circle.

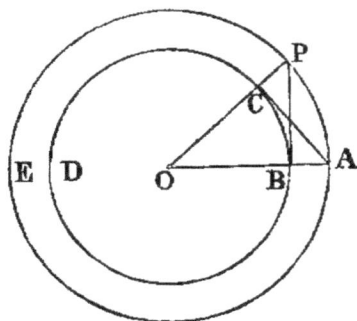

Fig. 1.

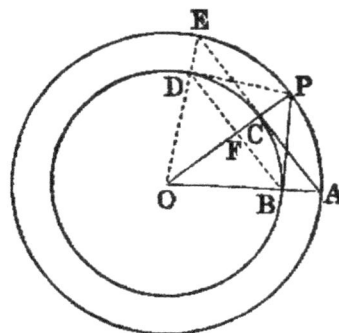

Fig. 2.

Sol.—Let O (fig. 1) be the centre of the given circle. Join OP, cutting the circumference in C. With O as centre, and OP as radius, describe the circle APE. Erect CA at right angles to OP. Join OA, intersecting the circle BCD in B. Join BP; *it will be the tangent required.*

Dem.—Since O is the centre of the two circles, we have OA equal to OP, and OC equal to OB. Hence the two triangles AOC, POB have the sides OA, OC in one respectively equal to the sides OP, OB in the other, and the contained angle common to both. Hence [I. IV.] the angle OCA is equal to OBP; but OCA is a right angle (const.); therefore OBP is a right angle, and [XVI.] *PB touches the circle at B.*

Cor.—If AC (fig. 2) be produced to E, OE joined, cutting the circle BCD in D, and the line DP drawn, *DP will be another tangent from P.*

Exercises.

1. The two tangents PB, PD (fig. 2) are equal to one another, because the square of each is equal to the square of OP minus the square of the radius.

2. If two circles be concentric, all tangents to the inner from points on the outer are equal.

3. If a quadrilateral be circumscribed to a circle, the sum of one pair of opposite sides is equal to the sum of the other pair.

4. If a parallelogram be circumscribed to a circle it must be a lozenge, and its diagonals intersect in the centre.

5. If BD be joined, intersecting OP in F, OP is perpendicular to BD.

6. The locus of the intersection of two equal tangents to two circles is a right line (*called the radical axis of the two circles*).

7. Find a point such that tangents from it to three given circles shall be equal. (*This point is called the radical centre of the three circles.*)

8. The rectangle $OF . OP$ is equal to the square of the radius.

DEF. *Two points, such as F and P, the rectangle of whose distances OF, OP from the centre is equal to the square of the radius, are called inverse points with respect to the circle.*

9. The intercept made on a variable tangent by two fixed tangents subtends a constant angle at the centre.

10. Draw a common tangent to two circles. Hence, show how to draw a line cutting two circles, so that the intercepted chords shall be of given lengths.

PROP. XVIII.—THEOREM

If a line (CD) touch a circle, the line (OC) from the centre to the point of contact is perpendicular to it.

Dem.—If not, suppose another line OG drawn from the centre to be perpendicular to CD. Let OG cut the circle in F. Then because the angle OGC is right (hyp.) the angle OCG [I. XVII.] must be acute. Therefore [I. XIX.] OC is greater than OG; but OC is equal to OF [I. Def. XXXII.]; therefore OF is greater than OG—that is, a part greater than the whole, which is impossible. *Hence OC must be perpendicular to CD.*

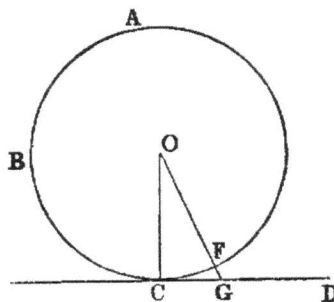

83

Or thus: Since the perpendicular must be the shortest line from O to CD, and OC is evidently the shortest line; therefore OC must be perpendicular to CD.

PROP. XIX.—THEOREM.

If a line (AB) be a tangent to a circle, the line (AC) drawn at right angles to it from the point of contact passes through the centre.

If the centre be not in AC, let O be the centre. Join AO. Then because AB touches the circle, and OA is drawn from the centre to the point of contact, OA is at right angles to AB [XVIII.]; therefore the angle OAB is right, and the angle CAB is right (hyp.); therefore OAB is equal to CAB—a part equal to the whole, which is impossible. *Hence the centre must be in the line AC.*

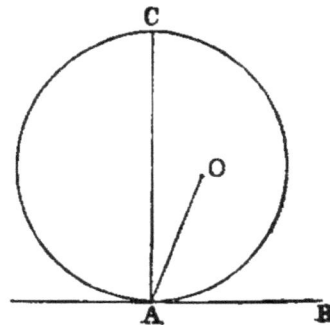

Cor.—If a number of circles touch the same line at the same point, the locus of their centres is the perpendicular to the line at the point.

Observation.—Propositions XVI., XVIII., XIX., are so related that any two can be inferred from the third by the "Rule of Identity." Hence it would, in strict logic, be sufficient to prove any one of the three, and the others would follow. Again, these three theorems are limiting cases of Proposition I., *Cor.* 1., and Parts 1, 2, of Proposition III., namely, when the points in which the chord cuts the circle become consecutive.

PROP. XX.—THEOREM.

The angle (AOB) at the centre (O) of a circle is double the angle (ACB) at the circumference standing on the same arc.

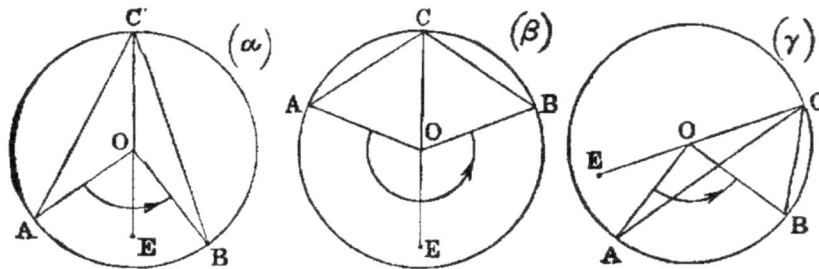

Dem.—Join CO, and produce it to E. Then because OA is equal to OC, the angle ACO is equal to OAC; but the angle AOE is equal to the sum of the two angles OAC, ACO. Hence the angle AOE is double the angle ACO. In like

84

manner the angle EOB is double the angle OCB. Hence (by adding in figs. (α), (β), and subtracting in (γ)), *the angle AOB is double of the angle ACB.*

Cor.—If AOB be a straight line, ACB will be a right angle—*that is, the angle in a semicircle is a right angle* (compare XXXI.).

PROP. XXI.—THEOREM.

The angles (ACB, ADB) in the same segment of a circle are equal.

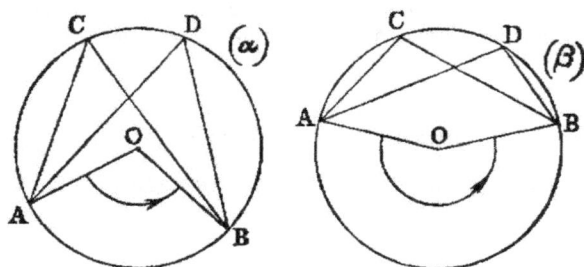

Dem.—Let O be the centre. Join OA, OB. Then the angle AOB is double of the angle ACB [XX.], and also double of the angle ADB. *Therefore the angle ACB is equal to the angle ADB.*

The following is the proof of the second part—that is, when the arc AB is not greater than a semicircle, without using angles greater than two right angles:—

Let O be the centre. Join CO, and produce it to meet the circle again in E. Join DE. Now since O is the centre, the segment ACE is greater than a semicircle; hence, by the first case, fig. (α), the angle ACE is equal to ADE. In like manner the angle ECB is equal to EDB. *Hence the whole angle ACB is equal to the whole angle ADB.*

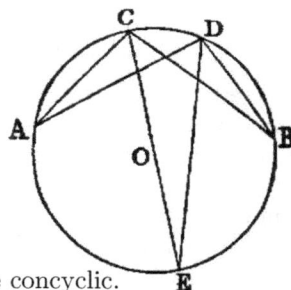

Cor. 1.—If two triangles ACB, ADB on the same base AB, and on the same side of it, have equal vertical angles, the four points A, C, D, B are concyclic.

Cor. 2.—If A, B be two fixed points, and if C varies its position in such a way that the angle ACB retains the same value throughout, the locus of C is a circle.

In other words—*Given the base of a triangle and the vertical angle, the locus of the vertex is a circle.*

85

1. Given the base of a triangle and the vertical angle, find the locus—
    (1) of the intersection of its perpendiculars;
    (2) of the intersection of the internal bisectors of its base angles;
    (3) of the intersection of the external bisectors of the base angles;
    (4) of the intersection of the external bisector of one base angle and the internal bisector of the other.

2. If the sum of the squares of two lines be given, their sum is a maximum when the lines are equal.

3. Of all triangles having the same base and vertical angle, the sum of the sides of an isosceles triangle is a maximum.

4. Of all triangles inscribed in a circle, the equilateral triangle has the maximum perimeter.

5. Of all concyclic figures having a given number of sides, the area is a maximum when the sides are equal.

PROP. XXII.—THEOREM.

The sum of the opposite angles of a quadrilateral (ABCD) inscribed in a circle is two right angles.

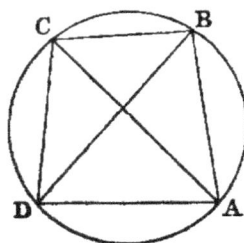 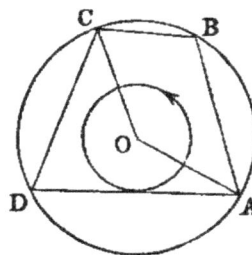

Fig. 1.   **Fig. 2.**

Dem.—Join AC, BD. The angle ABD is equal to ACD, being in the same segment $ABCD$ [XXI.]; and the angle DBC is equal to DAC, because they are in the same segment $DABC$. Hence the whole angle ABC is equal to the sum of the two angles ACD, DAC. To each add the angle CDA, and we have the sum of the two angles ABC, CDA equal to the sum of the three angles ACD, DAC, CDA of the triangle ACD; but the sum of the three angles of a triangle is equal to two right angles [I. XXXII.]. *Therefore the sum of ABC, CDA is two right angles.*

Or thus: Let O be the centre of the circle. Join OA, OC (*see* fig. 2). Now the angle AOC is double of CDA [XX.], and the angle COA is double of ABC. Hence the sum of the angles [I. Def. IX., note] AOC, COA is double of the sum of the angles CDA, ABC; but the sum of two angles AOC, COA is four right angles. *Therefore the sum of the angles CDA, ABC is two right angles.*

Or again: Let O be the centre (fig. 2). Join OA, OB OC, OD. Then the four triangles AOB, BOC, COD, DOA are each isosceles. Hence the angle OAB is equal to the angle OBA, and the angle OAD equal to the angle ODA; therefore the angle BAD is equal to the sum of the angles OBA, ODA. In like manner the angle BCD is equal to the sum of the angles OBC, ODC. Hence the sum of the two angles BAD, BCD is equal to the sum of the two angles ABC, ADC, *and hence each sum is two right angles.*

Cor.—If a parallelogram be inscribed in a circle it is a rectangle.

Exercises.

1. If the opposite angles of a quadrilateral be supplemental, it is cyclic.

2. If a figure of six sides be inscribed in a circle, the sum of any three alternate angles is four right angles.

3. A line which makes equal angles with one pair of opposite sides of a cyclic quadrilateral, makes equal angles with the remaining pair and with the diagonals.

4. If two opposite sides of a cyclic quadrilateral be produced to meet, and a perpendicular be let fall on the bisector of the angle between them from the point of intersection of the diagonals, this perpendicular will bisect the angle between the diagonals.

5. If two pairs of opposite sides of a cyclic hexagon be respectively parallel to each other, the remaining pair of sides are also parallel.

6. If two circles intersect in the points A, B, and any two lines ACD, BFE, be drawn through A and B, cutting one of the circles in the points C, E, and the other in the points D, F, the line CE is parallel to DF.

7. If equilateral triangles be described on the sides of any triangle, the lines joining the vertices of the original triangle to the opposite vertices of the equilateral triangles are concurrent.

8. In the same case prove that the centres of the circles described about the equilateral triangles form another equilateral triangle.

9. If a quadrilateral be described about a circle, the angles at the centre subtended by the opposite sides are supplemental.

10. The perpendiculars of a triangle are concurrent.

11. If a variable tangent meets two parallel tangents it subtends a right angle at the centre.

12. The feet of the perpendiculars let fall on the sides of a triangle from any point in the circumference of the circumscribed circle are collinear (SIMSON).

DEF.—*The line of collinearity is called Simson's line.*

13. If a hexagon be circumscribed about a circle, the sum of the angles subtended at the centre by any three alternate sides is equal to two right angles.

PROP. XXIII—THEOREM.

Two similar segments of circles which do not coincide cannot be constructed on the same chord (AB), and on the same side of that chord.

Dem.—If possible, let ACB, ADB, be two similar segments constructed on the same side of AB. Take any point D in the inner one. Join AD, and produce it to meet the outer one in C. Join BC, BD. Then since the segments are similar, the angle ADB is equal to ACB (Def. X.), which is impossible [I. XVI.]. *Hence two similar segments not coinciding cannot be described on the same chord and on the same side of it.*

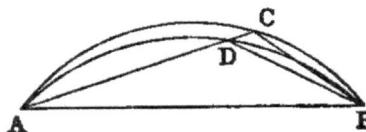

PROP. XXIV.—THEOREM.

Similar segments of circles (AEB, CFD) on equal chords (AB, CD) are equal to one another.

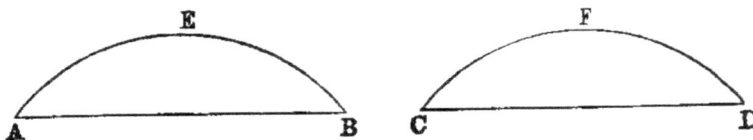

Dem.—Since the lines are equal, if AB be applied to CD, so that the point A will coincide with C, and the line AB with CD, the point B shall coincide with D; and because the segments are similar, they must coincide [XXIII.]. *Hence they are equal.*

This demonstration may be stated as follows:—Since the chords are equal, they are congruent; and therefore the segments, being similar, must be congruent.

PROP. XXV.—PROBLEM.

An arc (ABC) of a circle being given, it is required to describe the whole circle.

Sol.—Take any three points A, B, C in the arc. Join AB, BC. Bisect AB in D, and BC in E. Erect DF, EF at right angles to AB, BC; then F, the point of intersection, will be the centre of the circle.

Dem.—Because DF bisects the chord AB and is perpendicular to it, it passes through the centre [I., *Cor.* 1]. In like manner EF passes through the centre. *Hence the point F must be the centre; and the circle described from F as centre, with FA as radius, will be the circle required.*

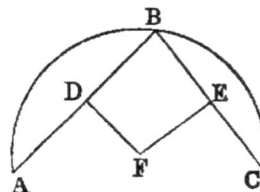

PROP. XXVI.—THEOREM.

The four Propositions XXVI.–XXIX. are so like in their enunciations that students frequently substitute one for another. The following scheme will assist in remembering them:—

In Proposition XXVI. are given angles =, to prove arcs =,
,, XXVII. ,, arcs =, ,, angles =,
,, XXVIII. ,, chords =, ,, arcs =,
,, XXIX. ,, arcs =, ,, chords =;

so that Proposition XXVII. is the converse of XXVI., and XXIX. of XXVIII.

In equal circles (ACB, DFE), equal angles at the centres (AOB, DHE) or at the circumferences (ACB, DFE) stand upon equal arcs.

Dem.—1. Suppose the angles at the centres to be given equal. Now because the circles are equal their radii are equal (Def. I.). Therefore the two triangles AOB, DHE have the sides AO, OB in one respectively equal to the sides DH, HE in the other, and the angle AOB equal to DHE (hyp.). Therefore [I. IV.] the base AB is equal to DE.

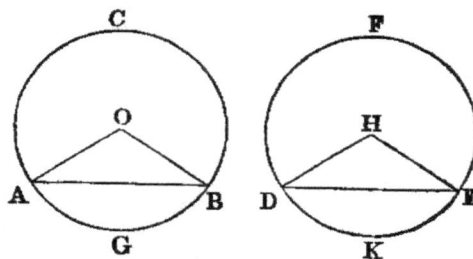

Again, since the angles ACB, DFE are [XX.] the halves of the equal angles AOB, DHE, they are equal [I. Axiom VII.]. Therefore (Def. X.) the segments ACB, DFE are similar, and their chords AB, DE have been proved equal; therefore [XXIV.] the segments are equal. And taking these equals from the whole circles, which are equal (hyp.), the remaining segments AGB, DKE are equal. *Hence the arcs AGB, DKE are equal.*

2. The demonstration of this case is included in the foregoing.

Cor. 1.—If the opposite angles of a cyclic quadrilateral be equal, one of its diagonals must be a diameter of the circumscribed circle.

Cor. 2.—Parallel chords in a circle intercept equal arcs.

Cor. 3.—If two chords intersect at any point within a circle, the sum of the opposite arcs which they intercept is equal to the arc which parallel chords intersecting on the circumference intercept. 2. If they intersect without the circle, the difference of the arcs they intercept is equal to the arc which parallel chords intersecting on the circumference intercept.

Cor. 4.—If two chords intersect at right angles, the sum of the opposite arcs which they intercept on the circle is a semicircle.

PROP. XXVII.—THEOREM.

In equal circles (ACB, DFE), angles at the centres (AOB, DHE), or at the circumferences (ACB, DFE), which stand on equal arcs (AB, DE), are equal.

Dem.—If possible let one of them, such as AOB, be greater than the other, DHE; and suppose a part such as AOL to be equal to DHE. Then since the circles are equal, and the angles AOL, DHE at the centres are equal (hyp.), the arc AL is equal to DE [XXVI.]; but AB is equal to DE (hyp.). Hence AL is equal to AB—that is, a part equal to the whole, which is absurd. *Therefore the angle AOB is equal to DHE.*

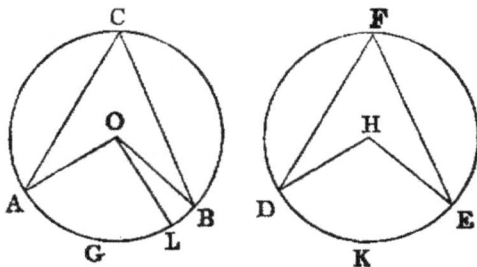

89

2. *The angles at the circumference, being the halves of the central angles, are therefore equal.*

PROP. XXVIII.—THEOREM.

In equal circles (ACB, DFE), equal chords (AB, DE) divide the circumferences into arcs, which are equal each to each— that is, the lesser to the lesser, and the greater to the greater.

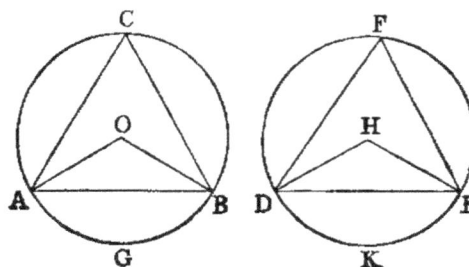

Dem.—If the equal chords be diameters, the Proposition is evident. If not, let O, H be the centres. Join AO, OB, DH, HE; then because the circles are equal their radii are equal (Def. I.). Hence the two triangles AOB, DHE have the sides AO, OB in one respectively equal to the sides DH, HE in the other, and the base AB is equal to DE (hyp.). Therefore [I. VIII.] the angle AOB is equal to DHE. *Hence the arc AGB is equal to DKE* [XXVI.]; and since the whole circumference AGBC is equal to the whole circumference DKEF, *the remaining arc ACB is equal to the remaining arc DFE.*

Exercises.

1. The line joining the feet of perpendiculars from any point in the circumference of a circle, on two diameters given in position, is given in magnitude.
2. If a line of given length slide between two lines given in position, the locus of the intersection of perpendiculars to the given lines at its extremities is a circle. (This is the converse of 1.)

PROP. XXIX.—THEOREM.

In equal circles (ACB, DFE), equal arcs (AGB, DCK) are subtended by equal chords.

Dem.—Let O, H be the centres (*see* last fig.). Join AO, OB, DH, HE; then because the circles are equal, the angles AOB, DHE at the centres, which stand on the equal arcs AGB, DKE, are equal [XXVII.]. Again, because the triangles AOB, DHE have the two sides AO, OB in one respectively equal to the two sides DH, HE in the other, and the angle AOB equal to the angle DHE, *the base AB of one is equal to the base DE of the other.*

Observation.—Since the two circles in the four last Propositions are equal, they are congruent figures, and the truth of the Propositions is evident by superposition.

90

PROP. XXX.—PROBLEM.

To bisect a given arc ACB.

Sol.—Draw the chord AB; bisect it in D; erect DC at right angles to AB, meeting the arc in C; then the arc is bisected in C.

Dem.—Join AC, BC. Then the triangles ADC, BDC have the side AD equal to DB (const.), and DC common to both, and the angle ADC equal to the angle BDC, each being right. Hence the base AC is equal to the base BC. Therefore [XXVIII.] the arc AC is equal to the arc BC. *Hence the arc AB is bisected in C.*

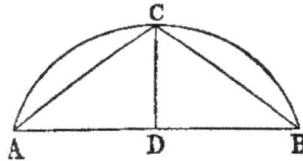

Exercises.

1. $ABCD$ is a semicircle whose diameter is AD; the chord BC produced meets AD produced in E: prove that if CE is equal to the radius, the arc AB is equal to three times CD.

2. The internal and the external bisectors of the vertical angle of a triangle inscribed in a circle meet the circumference again in points equidistant from the extremities of the base.

3. If from A, one of the points of intersection of two given circles, two chords ACD, $AC'D'$ be drawn, cutting the circles in the points C, D; C', D', the triangles BCD, $BC'D'$, formed by joining these to the second point B of intersection of the circles, are equiangular.

4. If the vertical angle ACB of a triangle inscribed in a circle be bisected by a line CD, which meets the circle again in D, and from D perpendiculars DE, DF be drawn to the sides, one of which must be produced: prove that EA is equal to BF, and hence show that CE is equal to half the sum of AC, BC.

PROP. XXXI.—THEOREM.

In a circle—(1). The angle in a semicircle is a right angle. (2). The angle in a segment greater than a semicircle is an acute angle. (3). The angle in a segment less than a semicircle is an obtuse angle.

Dem.—(1). Let AB be the diameter, C any point in the semicircle. Join AC, CB. *The angle ACB is a right angle.*

For let O be the centre. Join OC, and produce AC to F. Then because AO is equal to OC, the angle ACO is equal to the angle OAC. In like manner, the angle OCB is equal to CBO. Hence the angle ACB is equal to the sum of the two angles BAC, CBA; but [I. XXXII.] the angle FCB is equal to the sum of the two interior angles BAC, CBA of the triangle ABC. Hence the angle ACB is equal to its adjacent angle FCB, *and therefore it is a right angle* [I. Def. XIII.].

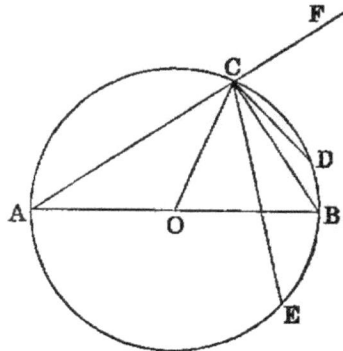

91

(2). Let the arc ACE be greater than a semicircle. Join CE. Then the angle ACE is evidently less than ACB; but ACB is right; *therefore ACE is acute.*

(3). Let the arc ACD be less than a semicircle; then evidently, from (1), *the angle ACD is obtuse.*

Cor. 1.—If a parallelogram be inscribed in a circle, its diagonals intersect at the centre of the circle.

Cor. 2.—Find the centre of a circle by means of a carpenter's square.

Cor. 3.—From a point outside a circle draw two tangents to the circle.

PROP. XXXII.—Theorem.

If a line (EF) be a tangent to a circle, and from the point of contact (A) a chord (AC) be drawn cutting the circle, the angles made by this line with the tangent are respectively equal to the angles in the alternate segments of the circle.

Dem.—(1). If the chord passes through the centre, the Proposition is evident, for the angles are right angles; but if not, from the point of contact A draw AB at right angles to the tangent. Join BC. Then because EF is a tangent to the circle, and AB is drawn from the point of contact perpendicular to EF, AB passes through this centre [XIX.]. Therefore the angle ACB is right [XXXI.]. Hence the sum of the two remaining angles ABC, CAB is one right angle; but the angle BAF is right (const.); therefore the sum of the angles ABC, BAC is equal to BAF. Reject BAC, which is common, and we get *the angle ABC equal to the angle FAC.*

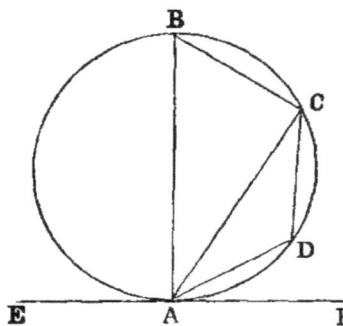

(2). Take any point D in the arc AC. It is required to prove that the angle CAE is equal to CDA.

Since the quadrilateral $ABCD$ is cyclic, the sum of the opposite angles ABC, CDA is two right angles [XXII.], and therefore equal to the sum of the angles FAC, CAE; but the angles ABC, FAC are equal (1). Reject them, and we get the *angle CDA equal to CAE.*

Or thus: Take any point G in the semicircle AGB. Join AG, GB, GC. Then the angle $AGB = FAB$, each being right, and $CGB = CAB$ [XXI.]. Therefore the remaining angle $AGC = FAC$. Again, join BD, CD. The angle $BDA = BAE$, each being right, and $CDB = CAB$ [XXI.]. *Hence the angle $CDA = CAE$.*—Lardner.

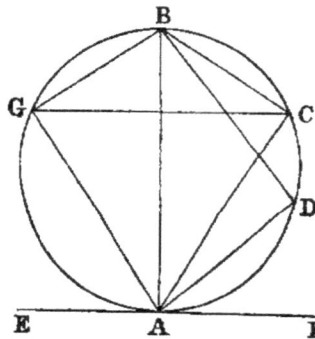

92

Or by the method of limits, see Townsend's *Modern Geometry*, vol. i., page 14.

The angle BAC is equal to BDC [XXI.]. Now let the point B move until it becomes consecutive to A; then AB will be a tangent, and BD will coincide with AD, and the angle BDC with ADC. Hence, if AX be a tangent at A, AC any chord, *the angle which the tangent makes with the chord is equal to the angle in the alternate segment.*

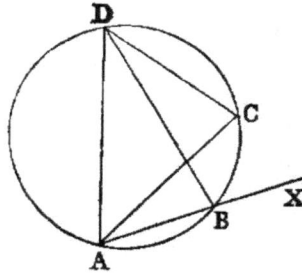

Exercises.

1. If two circles touch, any line drawn through the point of contact will cut off similar segments.

2. If two circles touch, and any two lines be drawn through the point of contact, cutting both circles again, the chord connecting their points of intersection with one circle is parallel to the chord connecting their points of intersection with the other circle.

3. ACB is an arc of a circle, CE a tangent at C, meeting the chord AB produced in E, and AD a perpendicular to AB in D: prove, if DE be bisected in C, that the arc $AC = 2CB$.

4. If two circles touch at a point A, and ABC be a chord through A, meeting the circles in B and C: prove that the tangents at B and C are parallel to each other, and that when one circle is within the other, the tangent at B meets the outer circle in two points equidistant from C.

5. If two circles touch externally, their common tangent at either side subtends a right angle at the point of contact, and its square is equal to the rectangle contained by their diameters.

PROP. XXXIII.—Problem.

On a given right line (AB) to describe a segment of a circle which shall contain an angle equal to a given rectilineal angle (X).

Sol.—If X be a right angle, describe a semicircle on the given line, and the thing required is done; for the angle in a semicircle is a right angle.

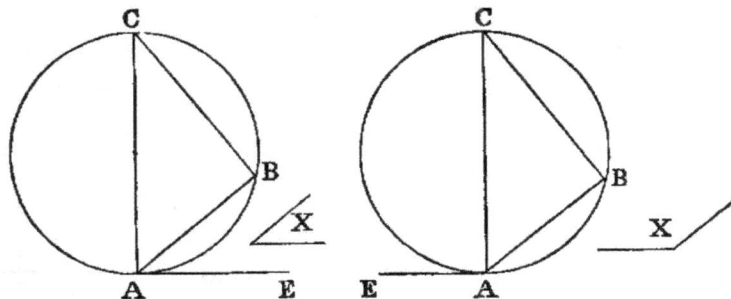

If not, make with the given line AB the angle BAE equal to X. Erect AC at right angles to AE, and BC at right angles to AB. On AC as diameter describe a circle: *it will be the circle required.*

93

Dem.—The circle on AC as diameter passes through B, since the angle ABC is right [XXXI.] and touches AE, since the angle CAE is right [XVI.]. Therefore the angle BAE [XXXII.] is equal to the angle in the alternate segment; but the angle BAE is equal to the angle X (const.). *Therefore the angle X is equal to the angle in the segment described on AB.*

Exercises.

1. Construct a triangle, being given base, vertical angle, and any of the following data:—1. Perpendicular. 2. The sum or difference of the sides. 3. Sum or difference of the squares of the sides. 4. Side of the inscribed square on the base. 5. The median that bisects the base.

2. If lines be drawn from a fixed point to all the points of the circumference of a given circle, the locus of all their points of bisection is a circle.

3. Given the base and vertical angle of a triangle, find the locus of the middle point of the line joining the vertices of equilateral triangles described on the sides.

4. In the same case, find the loci of the angular points of a square described on one of the sides.

PROP. XXXIV.—PROBLEM.

To cut off from a given circle (ABC) a segment which shall contain an angle equal to a given angle (X).

Sol.—Take any point A in the circumference. Draw the tangent AD, and make the angle DAC equal to the given angle X. AC will cut off the required segment.

Dem.—Take any point B in the alternate segment. Join BA, BC. Then the angle DAC is equal to ABC [XXXII.]; but DAC is equal to X (const.). *Therefore the angle ABC is equal to X.*

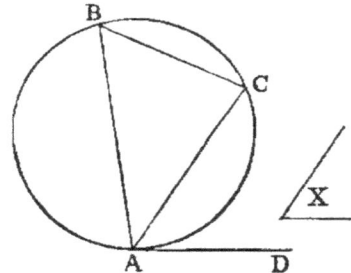

PROP. XXXV.—THEOREM.

If two chords (AB, CD) of a circle intersect in a point (E) within the circle, the rectangles ($AE.EB$, $CE.ED$) contained by the segments are equal.

Dem.—1. If the point of intersection be the centre, each rectangle is equal to the square of the radius. *Hence they are equal.*

2. Let one of the chords AB pass through the centre O, and cut the other chord CD, which does not pass through the centre, at right angles. Join OC. Now because AB passes through the centre, and cuts the other chord CD, which does not pass through the centre at right angles, it bisects it [III.]. Again, because AB is divided equally in O and unequally in E, the rectangle

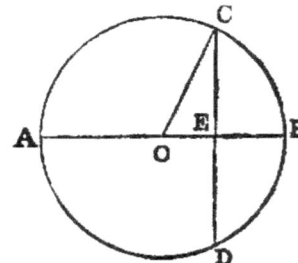

94

$AE \cdot EB$, together with OE^2, is equal to OB^2—that is, to OC^2 [II. v.]; but OC^2 is equal to $OE^2 + EC^2$ [I. xlvii.] Therefore

$$AE \cdot EB + OE^2 = OE^2 + EC^2.$$

Reject OE^2, which is common, and we have $AE \cdot EB = EC^2$; but CE^2 is equal to the rectangle $CE \cdot ED$, since CE is equal to ED. *Therefore the rectangle $AE \cdot EB$ is equal to the rectangle $CE \cdot ED$.*

3. Let AB pass through the centre, and cut CD, which does not pass through the centre obliquely. Let O be the centre. Draw OF perpendicular to CD [I. xi.]. Join OC, OD. Then, since CD is cut at right angles by OF, which passes through the centre, it is bisected in F [iii.], and divided unequally in E. Hence

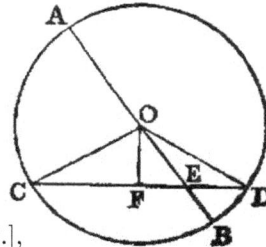

$$CE \cdot ED + FE^2 = FD^2 \text{ [II. v.]},$$
and
$$OF^2 = OF^2.$$

Hence, adding, since $FE^2 + OF^2 = OE^2$ [I. xlvii.], and $FD^2 + OF^2 = OD^2$, we get

$$CE \cdot ED + OE^2 = OD \text{ or } OB^2.$$

Again, since AB is bisected in O and divided unequally in E,

$$AE \cdot EB + OE^2 = OB^2 \text{ [II. v.]}.$$
Therefore
$$CE \cdot ED + OE^2 = AE \cdot EB + OE^2.$$
Hence
$$CE \cdot ED = AE \cdot EB.$$

4. Let neither chord pass through the centre. Through the point E, where they intersect, draw the diameter FG. Then by 3, the rectangle $FE \cdot EG$ is equal to the rectangle $AE \cdot EB$, and also to the rectangle $CE \cdot ED$. Hence the rectangle $AE \cdot EB$ is equal to the rectangle $CE \cdot ED$.

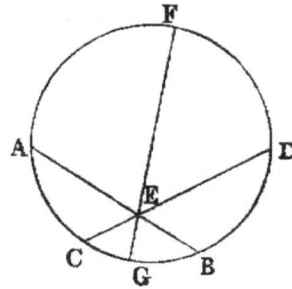

Cor. 1.—If a chord of a circle be divided in any point within the circle, the rectangle contained by its segments is equal to the difference between the square of the radius and the square of the line drawn from the centre to the point of section.

Cor. 2.—If the rectangle contained by the segments of one of two intersecting lines be equal to the rectangle contained by the segments of the other, the four extremities are concyclic.

Cor. 3.—*If two triangles be equiangular, the rectangle contained by the non-corresponding sides about any two equal angles are equal.*

95

Let *ABO*, *DCO* be the equiangular triangles, and let them be placed so that the equal angles at *O* may be vertically opposite, and that the non-corresponding sides *AO*, *CO* may be in one line; then the non-corresponding sides *BO*, *OD* shall be in one line. Now, since the angle *ABD* is equal to *ACD*, the points *A*, *B*, *C*, *D* are concyclic [XXI., *Cor*. 1]. *Hence the rectangle AO . OC is equal to the rectangle BO . OD* [XXXV.].

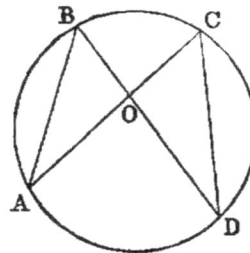

Exercises.

1. In any triangle, the rectangle contained by two sides is equal to the rectangle contained by the perpendicular on the third side and the diameter of the circumscribed circle.

Def.—*The supplement of an arc is the difference between it and a semicircle.*

2. The rectangle contained by the chord of an arc and the chord of its supplement is equal to the rectangle contained by the radius and the chord of twice the supplement.

3. If the base of a triangle be given, and the sum of the sides, the rectangle contained by the perpendiculars from the extremities of the base on the external bisector of the vertical angle is given.

4. If the base and the difference of the sides be given, the rectangle contained by the perpendiculars from the extremities of the base on the internal bisector is given.

5. Through one of the points of intersection of two circles draw a secant, so that the rectangle contained by the intercepted chords may be given, or a maximum.

6. *If the sum of two arcs, AC, CB of a circle be less than a semicircle, the rectangle AC . CB contained by their chords is equal to the rectangle contained by the radius, and the excess of the chord of the supplement of their difference above the chord of the supplement of their sum.*—CATALAN.

Dem.—Draw *DE*, the diameter which is perpendicular to *AB*, and draw the chords *CF*, *BG* parallel to *DE*. Now it is evident that the difference between the arcs *AC*, *CB* is equal to 2*CD*, and therefore = *CD* + *EF*. Hence the arc *CBF* is the supplement of the difference, and *CF* is the chord of that supplement. Again, since the angle *ABG* is right, the arc *ABG* is a semicircle. Hence *BG* is the supplement of the sum of the arcs *AC*, *CB*; therefore the line *BG* is the chord of the supplement of the sum. Now (Ex. 1), the rectangle *AC* . *CB* is equal to the rectangle contained by the diameter and *CI*, and therefore equal to the rectangle contained by the radius and 2*CI*; but the difference between *CF* and *BG* is evidently equal to 2*CI*. *Hence the rectangle AC . CB is equal to the rectangle contained by the radius and the difference between the chords CF, BG.*

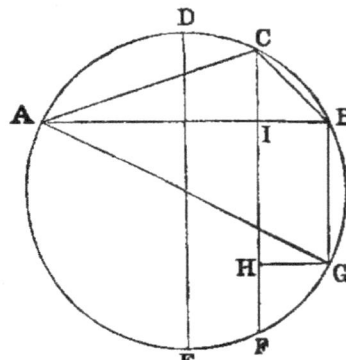

7. If we join *AF*, *BF* we find, as before, the rectangle *AF* . *FB* equal to the rectangle contained by the radius and 2*FI*—that is, equal to the rectangle contained by the radius and the sum of *CF* and *BG*. Hence—*If the sum of two arcs of a circle be greater than a semicircle, the rectangle contained by their chords is equal to the rectangle contained by the radius, and the sum of the chords of the supplements of their sum and their difference.*

8. Through a given point draw a transversal cutting two lines given in position, so that the rectangle contained by the segments intercepted between it and the line may be given.

PROP. XXXVI.—THEOREM.

If from any point (P) without a circle two lines be drawn to it, one of which (PT) is a tangent, and the other (PA) a secant, the rectangle (AP, BP) contained by the segments of the secant is equal to the square of the tangent.

Dem.—1. Let PA pass through the centre O. Join OT. Then because AB is bisected in O and divided externally in P, the rectangle $AP . BP + OB^2$ is equal to OP^2 [II. VI.]. But since PT is a tangent, and OT drawn from the centre to the point of contact, the angle OTP is right [XVIII.]. Hence $OT^2 + PT^2$ is equal to OP^2.

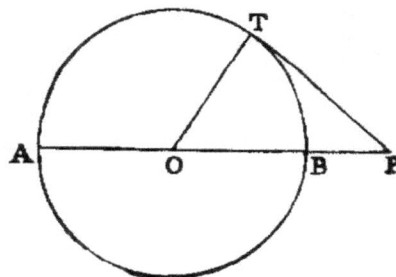

Therefore $\qquad AP . BP + OB^2 = OT^2 + PT^2;$

but $\qquad\qquad\qquad OB^2 = OT^2.$

Hence the rectangle $\qquad AP . BP = PT^2.$

2. If AB does not pass through the centre O, let fall the perpendicular OC on AB. Join OT, OB, OP. Then because OC, a line through the centre, cuts AB, which does not pass through the centre at right angles, it bisects it [III.]. Hence, since AB is bisected in C and divided externally in P, the rectangle

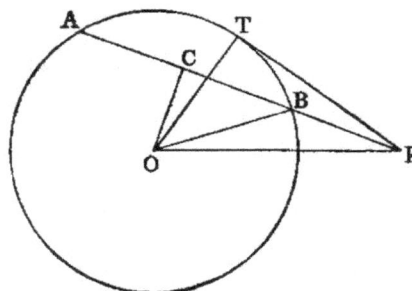

$$AP . BP + CB^2 = CP^2 \text{ [II. VI.];}$$

and $\qquad\qquad\qquad OC^2 = OC^2.$

Hence, adding, since $CB^2 + OC^2 = OB^2$ [I. XLVII.], and $CP^2 + OC^2 = OP^2$, we get

rectangle $\qquad\qquad AP . BP + OB^2 = OP^2;$

but $\qquad\qquad\qquad OT^2 + PT^2 = OP^2$ [I. XLVII.].

Therefore $\qquad AP . BP + OB^2 = OT^2 + PT^2;$

and rejecting the equals OB^2 and OT^2, we have the rectangle

$$AP . BP = PT^2.$$

The two Propositions XXXV., XXXVI., may be included in one enunciation, as follows:—*The rectangle AP . BP contained by the segments of any chord of a given circle passing through a fixed point P, either within or without the circle, is constant. For let O be the centre: join*

OA, OB, OP. Then OAB is an isosceles triangle, and OP is a line drawn from its vertex to a point P in the base, or base produced. Then the rectangle $AP \cdot BP$ is equal to the difference of the squares of OB and OP, and is therefore constant.

Cor. 1.—If two lines AB, CD produced meet in P, and if the rectangle $AP \cdot BP = CP \cdot DP$, the points A, B, C, D are concyclic (compare XXXV., *Cor.* 2).

Cor. 2.—Tangents to two circles from any point in their common chord are equal (compare XVII., Ex. 6).

Cor. 3.—The common chords of any three intersecting circles are concurrent (compare XVII., Ex. 7).

Exercise.

If from the vertex A of a $\triangle ABC$, AD be drawn, meeting CB produced in D, and making the angle $BAD = ACB$, prove $DB \cdot DC = DA^2$.

PROP. XXXVII.—THEOREM.

If the rectangle $(AP \cdot BP)$ contained by the segments of a secant, drawn from any point (P) without a circle, be equal to the square of a line (PT) drawn from the same point to meet the circle, the line which meets the circle is a tangent.

Dem.—From P draw PQ touching the circle [XVII.]. Let O be the centre. Join OP, OQ, OT. Now the rectangle $AP \cdot BP$ is equal to the square on PT (hyp.), and equal to the square on PQ [XXXVI.]. Hence PT^2 is equal to PQ^2, and therefore PT is equal to PQ. Again, the triangles OTP, OQP have the side OT equal OQ, TP equal QP, and the base OP common; hence [I. VIII.] the angle OTP is equal to OQP; but OQP is a right angle, since PQ is a tangent [XVIII.]; hence OTP is right, *and therefore* [XVI.] *PT is a tangent.*

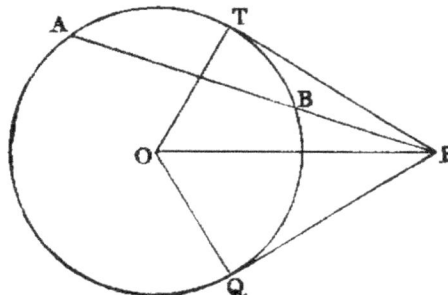

Exercises.

1. Describe a circle passing through two given points, and fulfilling either of the following conditions: 1, touching a given line; 2, touching a given circle.

2. Describe a circle through a given point, and touching two given lines; or touching a given file and a given circle.

3. Describe a circle passing through a given point, having its centre on a given line and touching a given circle.

4. Describe a circle through two given points, and intercepting a given arc on a given circle.

5. A, B, C, D are four collinear points, and EF is a common tangent to the circles described upon AB, CD as diameters: prove that the triangles AEB, CFD are equiangular.

6. The diameter of the circle inscribed in a right-angled triangle is equal to half the sum of the diameters of the circles touching the hypotenuse, the perpendicular from the right angle of the hypotenuse, and the circle described about the right-angled triangle.

98

Questions for Examination on Book III.

1. What is the subject-matter of Book III.?
2. Define equal circles.
3. What is the difference between a chord and a secant?
4. When does a secant become a tangent?
5. What is the difference between a segment of a circle and a sector?
6. What is meant by an angle in a segment?
7. If an arc of a circle be one-sixth of the whole circumference, what is the magnitude of the angle in it?
8. What are linear segments?
9. What is meant by an angle standing on a segment?
10. What are concyclic points?
11. What is a cyclic quadrilateral?
12. How many intersections can a line and a circle have?
13. What does the line become when the points of intersection become consecutive?
14. How many points of intersection can two circles have?
15. What is the reason that if two circles touch they cannot have any other common point?
16. Give one enunciation that will include Propositions XI., XII. of Book III.
17. What Proposition is this a limiting case of?
18. Explain the extended meaning of the word angle.
19. What is Euclid's limit of an angle?
20. State the relations between Propositions XVI., XVIII., XIX.
21. What Propositions are these limiting cases of?
22. How many common tangents can two circles have?
23. What is the magnitude of the rectangle of the segments of a chord drawn through a point 3.65 metres distant from the centre of a circle whose radius is 4.25 metres?
24. The radii of two circles are 4.25 and 1.75 feet respectively, and the distance between their centres 6.5 feet; find the lengths of their direct and their transverse common tangents.
25. If a point be h feet outside the circumference of a circle whose diameter is 7920 miles, prove that the length of the tangent drawn from it to the circle is $\sqrt{\dfrac{3h}{2}}$ miles.
26. Two parallel chords of a circle are 12 perches and 16 perches respectively, and their distance asunder is 2 perches; find the length of the diameter.
27. What is the locus of the centres of all circles touching a given circle in a given point?
28. What is the condition that must be fulfilled that four points may be concyclic?
29. If the angle in a segment of a circle be a right angle and a-half, what part of the whole circumference is it?
30. Mention the converse Propositions of Book III. which are proved directly.
31. What is the locus of the middle points of equal chords in a circle?
32. The radii of two circles are 6 and 8, and the distance between their centres 10; find the length of their common chord.
33. If a figure of any even number of sides be inscribed in a circle, prove that the sum of one set of alternate angles is equal to the sum of the remaining angles.

Exercises on Book III.

1. If two chords of a circle intersect at right angles, the sum of the squares on their segments is equal to the square on the diameter.
2. If a chord of a given circle subtend a right angle at a fixed point, the rectangle of the perpendiculars on it from the fixed point and from the centre of the given circle is constant. Also the sum of the squares of perpendiculars on it from two other fixed points (which may be found) is constant.
3. If through either of the points of intersection of two equal circles any line be drawn meeting them again in two points, these points are equally distant from the other intersection of the circles.

99

4. Draw a tangent to a given circle so that the triangle formed by it and two fixed tangents to the circle shall be—1, a maximum; 2, a minimum.

5. If through the points of intersection A, B of two circles any two lines ACD, BEF be drawn parallel to each other, and meeting the circles again in C, D, E, F; then $CD = EF$.

6. In every triangle the bisector of the greatest angle is the least of the three bisectors of the angles.

7. The circles whose diameters are the four sides of any cyclic quadrilateral intersect again in four concyclic points.

8. The four angular points of a cyclic quadrilateral determine four triangles whose ortho-centres (the intersections of their perpendiculars) form an equal quadrilateral.

9. If through one of the points of intersection of two circles we draw two common chords, the lines joining the extremities of these chords make a given angle with each other.

10. The square on the perpendicular from any point in the circumference of a circle, on the chord of contact of two tangents, is equal to the rectangle of the perpendiculars from the same point on the tangents.

11. Find a point in the circumference of a given circle, the sum of the squares on whose distances from two given points may be a maximum or a minimum.

12. Four circles are described on the sides of a quadrilateral as diameters. The common chord of any two on adjacent sides is parallel to the common chord of the remaining two.

13. The rectangle contained by the perpendiculars from any point in a circle, on the diagonals of an inscribed quadrilateral, is equal to the rectangle contained by the perpendiculars from the same point on either pair of opposite sides.

14. The rectangle contained by the sides of a triangle is greater than the square on the internal bisector of the vertical angle, by the rectangle contained by the segments of the base.

15. If through A, one of the points of intersection of two circles, we draw any line ABC, cutting the circles again in B and C, the tangents at B and C intersect at a given angle.

16. If a chord of a given circle pass through a given point, the locus of the intersection of tangents at its extremities is a right line.

17. The rectangle contained by the distances of the point where the internal bisector of the vertical angle meets the base, and the point where the perpendicular from the vertex meets it from the middle point of the base, is equal to the square on half the difference of the sides.

18. State and prove the Proposition analogous to 17 for the external bisector of the vertical angle.

19. The square on the external diagonal of a cyclic quadrilateral is equal to the sum of the squares on the tangents from its extremities to the circumscribed circle.

20. If a variable circle touch a given circle and a given line, the chord of contact passes through a given point.

21. If A, B, C be three points in the circumference of a circle, and D, E the middle points of the arcs AB, AC; then if the line DE intersect the chords AB, AC in the points F, G, AF is equal to AG.

22. Given two circles, O, O'; then if any secant cut O in the points B, C, and O' in the points B', C', and another secant cuts them in the points D, E; D', E' respectively; the four chords BD, CE, $B'D'$, $C'E'$ form a cyclic quadrilateral.

23. If a cyclic quadrilateral be such that a circle can be inscribed in it, the lines joining the points of contact are perpendicular to each other.

24. If through the point of intersection of the diagonals of a cyclic quadrilateral the minimum chord be drawn, that point will bisect the part of the chord between the opposite sides of the quadrilateral.

25. Given the base of a triangle, the vertical angle, and either the internal or the external bisector at the vertical angle; construct it.

26. If through the middle point A of a given arc BAC we draw any chord AD, cutting BC in E, the rectangle $AD \cdot AE$ is constant.

27. The four circles circumscribing the four triangles formed by any four lines pass through a common point.

28. If X, Y, Z be any three points on the three sides of a triangle ABC, the three circles about the triangles YAZ, ZBX, XCY pass through a common point.

29. If the position of the common point in the last question be given, the three angles of the triangle XYZ are given, and conversely.

30. Place a given triangle so that its three sides shall pass through three given points.

31. Place a given triangle so that its three vertices shall lie on three given lines.

32. Construct the greatest triangle equiangular to a given one whose sides shall pass through three given points.

33. Construct the least triangle equiangular to a given one whose vertices shall lie on three given lines.

34. Construct the greatest triangle equiangular to a given one whose sides shall touch three given circles.

35. If two sides of a given triangle pass through fixed points, the third touches a fixed circle.

36. If two sides of a given triangle touch fixed circles, the third touches a fixed circle.

37. Construct an equilateral triangle having its vertex at a given point, and the extremities of its base on a given circle.

38. Construct an equilateral triangle having its vertex at a given point, and the extremities of its base on two given circles.

39. Place a given triangle so that its three sides shall touch three given circles.

40. Circumscribe a square about a given quadrilateral.

41. Inscribe a square in a given quadrilateral.

42. Describe circles—(1) orthogonal (cutting at right angles) to a given circle and passing through two given points; (2) orthogonal to two others, and passing through a given point; (3) orthogonal to three others.

43. If from the extremities of a diameter AB of a semicircle two chords AD, BE be drawn, meeting in C, $AC \cdot AD + BC \cdot BE = AB^2$.

44. If $ABCD$ be a cyclic quadrilateral, and if we describe any circle passing through the points A and B, another through B and C, a third through C and D, and a fourth through D and A; these circles intersect successively in four other points E, F, G, H, forming another cyclic quadrilateral.

45. If ABC be an equilateral triangle, what is the locus of the point M, if $MA = MB + MC$?

46. In a triangle, given the sum or the difference of two sides and the angle formed by these sides both in magnitude and position, the locus of the centre of the circumscribed circle is a right line.

47. Describe a circle—(1) through two given points which shall bisect the circumference of a given circle; (2) through one given point which shall bisect the circumference of two given circles.

48. Find the locus of the centre of a circle which bisects the circumferences of two given circles.

49. Describe a circle which shall bisect the circumferences of three given circles.

50. AB is a diameter of a circle; AC, AD are two chords meeting the tangent at B in the points E, F respectively: prove that the points C, D, E, F are concyclic.

51. CD is a perpendicular from any point C in a semicircle on the diameter AB; EFG is a circle touching DB in E, CD in F, and the semicircle in G; prove—(1) that the points A, F, G are collinear; (2) that $AC = AE$.

52. Being given an obtuse-angled triangle, draw from the obtuse angle to the opposite side a line whose square shall be equal to the rectangle contained by the segments into which it divides the opposite side.

53. O is a point outside a circle whose centre is E; two perpendicular lines passing through O intercept chords AB, CD on the circle; then $AB^2 + CD^2 + 4OE^2 = 8R^2$.

54. The sum of the squares on the sides of a triangle is equal to twice the sum of the rectangles contained by each perpendicular and the portion of it comprised between the corresponding vertex and the orthocentre; also equal to $12R^2$ minus the sum of the squares of the distances of the orthocentre from the vertices.

55. If two circles touch in C, and if D be any point outside the circles at which their radii through C subtend equal angles, if DE, DF be tangent from D, $DE \cdot DF = DC^2$.

101

BOOK IV.

INSCRIPTION AND CIRCUMSCRIPTION OF TRIANGLES AND OF REGULAR POLYGONS IN AND ABOUT CIRCLES

DEFINITIONS.

I. If two rectilineal figures be so related that the angular points of one lie on the sides of the other—1, the former is said to be inscribed in the latter; 2, the latter is said to be described about the former.

II. A rectilineal figure is said to be inscribed in a circle when its angular points are on the circumference. *Reciprocally*, a rectilineal figure is said to be circumscribed to a circle when each side touches the circle.

III. A circle is said to be inscribed in a rectilineal figure when it touches each side of the figure. *Reciprocally*, a circle is said to be circumscribed to a rectilineal figure when it passes through each angular point of the figure.

IV. A rectilineal figure which is both equilateral and equiangular is said to be regular.

Observation.—The following summary of the contents of the Fourth Book will assist the student in remembering it:—

1. It contains sixteen Propositions, of which four relate to triangles, four to squares, four to pentagons, and four miscellaneous Propositions.
2. Of the four Propositions occupied with triangles—
(α) One is to inscribe a triangle in a circle.
(β) Its reciprocal, to describe a triangle about a circle.
(γ) To inscribe a circle in a triangle.
(δ) Its reciprocal, to describe a circle about a triangle.
3. If we substitute in (α), (β), (γ), (δ) squares for triangles, and pentagons for triangles, we have the problems for squares and pentagons respectively.
4. Every Proposition in the fourth Book is a problem.

PROP. I.—Problem.

In a given circle (ABC) to place a chord equal to a given line (D) not greater than the diameter.

Sol.—Draw any diameter AC of the circle; then, if AC be equal to D, the thing required is done; if not, from AC cut off the part AE equal to D [I. III.]; and with A as centre and AE as radius, describe the circle EBF, cutting the circle ABC in the points B, F. Join AB. *Then AB is the chord required.*

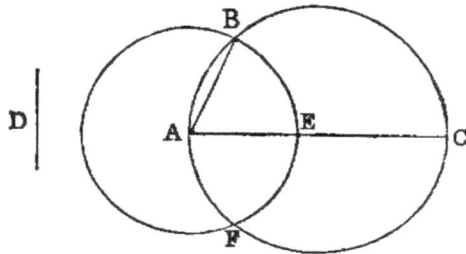

Dem.—Because A is the centre of the circle EBF, AB is equal to AE [I. Def. XXXII.]; but AE is equal to D (const.); *therefore AB is equal to D.*

PROP. II.—PROBLEM.

In a given circle (ABC) to inscribe a triangle equiangular to a given triangle (DEF).

Sol.—Take any point A in the circumference, and at it draw the tangent GH; then make the angle HAC equal to E, and GAB equal to F [I. XXIII.] Join BC. *ABC is a triangle fulfilling the required conditions.*

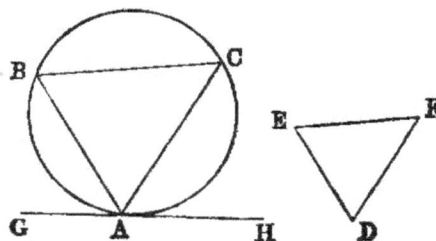

Dem.—The angle E is equal to HAC (const.), and HAC is equal to the angle ABC in the alternate segment [III. XXXII.]. Hence the angle E is equal to ABC. In like manner the angle F is equal to ACB. Therefore [I. XXXII.] the remaining angle D is equal to BAC. *Hence the triangle ABC inscribed in the given circle is equiangular to DEF.*

PROP. III.—PROBLEM.

About a given circle (ABC) to describe a triangle equiangular to a given triangle (DEF).

Sol.—Produce any side DE of the given triangle both ways to G and H, and from the centre O of the circle draw any radius OA; make the angle AOB equal to GEF [I. XXIII.], and the angle AOC equal to HDF. At the points A, B, C draw the tangents LM, MN, NL to the given circle. *LMN is a triangle fulfilling the required conditions.*

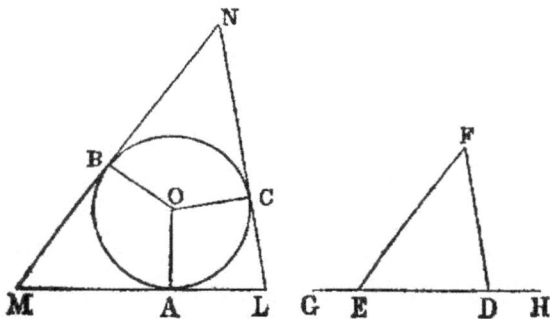

103

Dem.—Because AM touches the circle at A, the angle OAM is right. In like manner, the angle MBO is right; but the sum of the four angles of the quadrilateral $OAMB$ is equal to four right angles. Therefore the sum of the two remaining angles AOB, AMB is two right angles; and [I. XIII.] the sum of the two angles GEF, FED is two right angles. Therefore the sum of AOB, AMB is equal to the sum of GEF, FED; but AOB is equal to GEF (const.). Hence AMB is equal to FED. In like manner, ALC is equal to EDF; therefore [I. XXXII.] the remaining angle BNC is equal to DFE. *Hence the triangle LMN is equiangular to DEF.*

PROP. IV.—PROBLEM.

To inscribe a circle in a given triangle (ABC).

Sol.—Bisect any two angles A, B of the given triangle by the lines AO, BO; *then O, their point of intersection, is the centre of the required circle.*

Dem.—From O let fall the perpendiculars OD, OE, OF on the sides of the triangle. Now, in the triangles OAE, OAF the angle OAE is equal to OAF (const.), and the angle AEO equal to AFO, because each is right, and the side OA common. Hence [I. XXVI.] the side OE is equal to OF. In like manner OD is equal to OF; therefore the three lines OD, OE, OF are all equal. And the circle described with O as centre and OD as radius will pass through the points E, F; and since the angles D, E, F are right, it will [III. XVI.] touch the three sides of the triangle ABC; *and therefore the circle DEF is inscribed in the triangle ABC.*

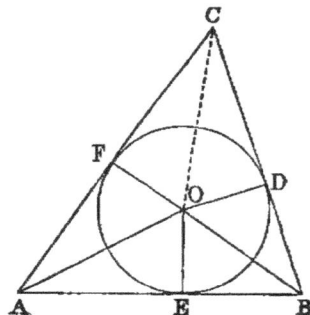

Exercises.

1. If the points O, C be joined, the angle C is bisected. Hence "the bisectors of the angles of a triangle are concurrent" (compare I. XXVI., Ex. 7).

2. If the sides BC, CA, AB of the triangle ABC be denoted by a, b, c, and half their sum by s, the distances of the vertices A, B, C of the triangle from the points of contact of the inscribed circle are respectively $s - a$, $s - b$, $s - c$.

3. If the external angles of the triangle ABC be bisected as in the annexed diagram, the three angular points O', O'', O''', of the triangle formed by the three bisectors will be the centres of three circles, each touching one side externally, and the other two produced. These three circles are called the *escribed* circles of the triangle ABC.

4. The distances of the vertices A, B, C from the points of contact of the escribed circle which touches AB externally are $s - b$, $s - a$, s.

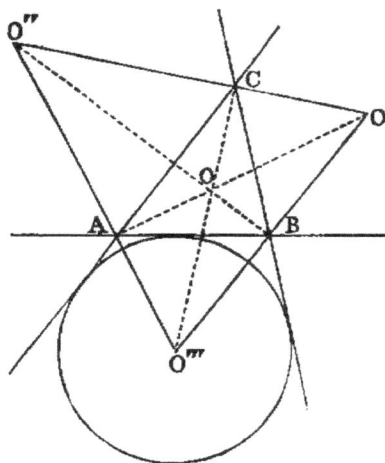

5. The centre of the inscribed circle, the centre of each escribed circle, and two of the angular points of the triangle, are concyclic. Also any two of the escribed centres are concyclic with the corresponding two of the angular points of the triangle.

6. Of the four points O, O', O'', O''', any one is the orthocentre of the triangle formed by the remaining three.

7. The three triangles BCO', CAO'', ABO''' are equiangular.

8. The rectangle $CO \cdot CO''' = ab$; $AO \cdot AO' = bc$; $BO \cdot BO'' = ca$.

9. Since the whole triangle ABC is made up of the three triangles AOB, BOC, COA, we see that the rectangle contained by the sum of the three sides, and the radius of the inscribed circle, is equal to twice the area of the triangle. Hence, if r denote the radius of the inscribed circle, $rs = $ area of the triangle.

10. If r' denote the radius of the escribed circle which touches the side a externally, it may be shown in like manner that $r'(s-a) = $ area of the triangle.

11. $rr' = s - b \cdot s - c$.

12. Square of area $= s \cdot s - a \cdot s - b \cdot s - c$.

13. Square of area $= r \cdot r' \cdot r'' \cdot r'''$.

14. If the triangle ABC be right-angled, having the angle C right,

$$r = s - c; \ r' = s - b; \ r'' = s - a; \ r''' = s.$$

15. Given the base of a triangle, the vertical angle, and the radius of the inscribed, or any of the escribed circles: construct it.

PROP. V.—PROBLEM.

To describe a circle about a given triangle (ABC).

Sol.—Bisect any two sides BC, AC in the points D, E. Erect DO, EO at right angles to BC, CA; *then O, the point of intersection of the perpendiculars, is the centre of the required circle.*

Dem.—Join OA, OB, OC. The triangles BDO, CDO have the side BD equal CD (const.), and DO common, and the angle BDO equal to the angle CDO, because each is right. Hence [I. IV.] BO is equal to OC. In like manner AO is equal to OC. Therefore the three lines AO, BO, CO are equal, and the circle described with O as centre, and OA as radius, will pass through the points A, B, C, *and be described about the triangle ABC.*

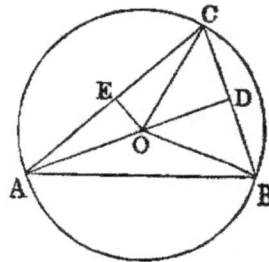

Cor. 1.—Since the perpendicular from O on AB bisects it [III. III.], we see that the perpendiculars at the middle points of the sides of a triangle are concurrent.

DEF.—*The circle ABC is called the circumcircle, its radius the circumradius, and its centre the circumcentre of the triangle.*

Exercises.

1. The three perpendiculars of a triangle (ABC) are concurrent.

Dem.—Describe a circle about the triangle. Let fall the perpendicular CF. Produce CF to meet the circle in G. Make $FO = FG$. Join AG, AO. Produce AO to meet BC in D. Then the triangles GFA, OFA have the sides GF, FA in one equal to the sides OF, FA in the other, and the contained angles equal. Hence [I. IV.] the angle GAF equal OAF; but $GAF = GCB$ [III. XXI.]; hence $OAF = OCD$, and $FOA = DOC$; hence $OFA = ODC$; but OFA is right, hence ODC is right. In like manner, if BO be joined to meet AC in E, BE will be perpendicular to AC. *Hence the three perpendiculars pass through O, and are concurrent.* This Proposition may be proved simply as follows:—

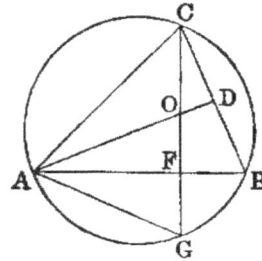

Draw parallels to the sides of the original triangle ABC through its vertices, forming a new triangle $A'B'C'$ described about ABC; then the three perpendiculars at the middle points of the sides of $A'B'C'$ are concurrent [V. *Cor.* 1], and these are evidently the perpendiculars from the vertices on the opposite sides of the triangle ABC (compare Ex. 16, Book I.).

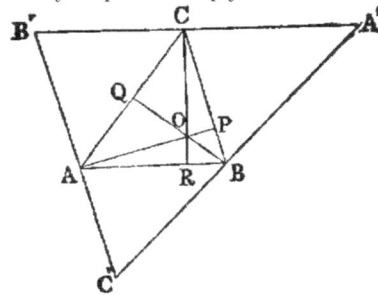

Def.—*The point O is called the orthocentre of the triangle ABC.*

2. The three rectangles $OA \cdot OP$, $OB \cdot OQ$, $OC \cdot OR$ are equal.

Def.—*The circle round O as centre, the square of whose radius is equal $OA \cdot OP = OB \cdot OQ = OC \cdot OR$, is called the polar circle of the triangle ABC.*

Observation.—If the orthocentre of the triangle ABC be within the triangle, the rectangles $OA \cdot OP$, $OB \cdot OQ$, $OC \cdot OR$ are negative, because the lines $OA \cdot OP$, &c., are measured in opposite directions, and have contrary signs; hence the *polar circle* is imaginary; but it is real when the point O is without the triangle—that is, when the triangle has an obtuse angle.

3. If the perpendiculars of a triangle be produced to meet the circumscribed circle, the intercepts between the orthocentre and the circle are bisected by the sides of the triangle.

4. The point of bisection (I) of the line (OP) joining the orthocentre (O) to the circumference (P) of any triangle is equally distant from the feet of the perpendiculars, from the middle points of the sides, and from the middle points of the distances of the vertices from the orthocentre.

Dem.—Draw the perpendicular PH; then, since OF, PH are perpendiculars on AB, and OP is bisected in I, it is easy to see that $IH = IF$. Again, since OP, OG are bisected in I, F; $IF = \frac{1}{2}PG$— that is, $IF = \frac{1}{2}$ the radius. Hence the distance of I from the foot of each perpendicular, and from the middle point of each side, is $= \frac{1}{2}$ the radius. In like manner, if OC be bisected in K, then $IK = \frac{1}{2}$ the radius. Hence we have the following theorem:—*The nine points made up of the feet of the perpendiculars, the middle points of the sides, and the middle points of the lines from the vertices to the orthocentre, are concyclic.*

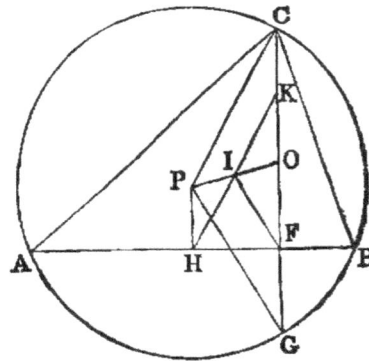

Def.—*The circle through these nine points is called the "nine points circle" of the triangle.*

5. The circumcircle of a triangle is the "nine points circle" of each of the four triangles formed by joining the centres of the inscribed and escribed circles.

6. The distances between the vertices of a triangle and its orthocentre are respectively the doubles of the perpendiculars from the circumcentre on the sides.

7. The radius of the "nine points circle" of a triangle is equal to half its circumradius.

106

PROP. VI.—PROBLEM.

In a given circle (ABCD) to inscribe a square.

Sol.—Draw any two diameters AC, BD at right angles to each other. Join AB, BC, CD, DA. *ABCD is a square.*

Dem.—Let O be the centre. Then the four angles at O, being right angles, are equal. Hence the arcs on which they stand are equal [III. XXVI.], and hence the four chords are equal [III. XXIX.]. Therefore the figure $ABCD$ is equilateral.

Again, because AC is a diameter, the angle ABC is right [III. XXXI.]. In like manner the remaining angles are right. *Hence ABCD is a square.*

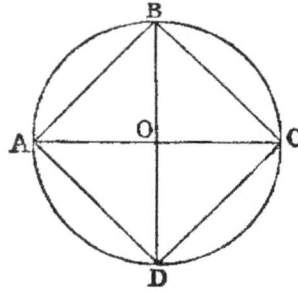

PROP. VII.—PROBLEM.

About a given circle (ABCD) to describe a square.

Sol.—Through the centre O draw any two diameters at right angles to each other, and draw at the points A, B, C, D the lines HE, EF, FG, GH touching the circle. *EFGH is a square.*

Dem.—Because AE touches the circle at A, the angle EAO is right [III. XVIII.], and therefore equal to BOC, which is right (const.). Hence AE is parallel to OB. In like manner EB is parallel to AO; and since AO is equal to OB, the figure $AOBE$ is a lozenge, and the angle AOB is right; hence $AOBE$ is a square. In like manner each of the figures BC, CD, DA is a square. *Hence the whole figure is a square.*

Cor.—The circumscribed square is double of the inscribed square.

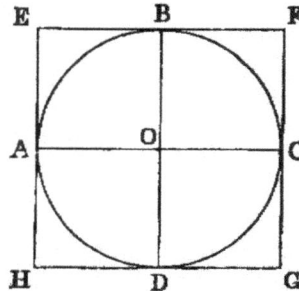

PROP. VIII.—PROBLEM.

In a given square (ABCD) to inscribe a circle.

Sol.—Bisect (*see* last diagram) two adjacent sides EH, EF in the points A, B, and through A, B draw the lines AC, BD, respectively parallel to EF, EH; *then O, the point of intersection of these parallels, is the centre of the required circle.*

Dem.—Because $AOBE$ is a parallelogram, its opposite sides are equal; therefore AO is equal to EB; but EB is half the side of the given square;

107

therefore AO is equal to half the side of the given square; and so in like manner is each of the lines OB, OC, OD; therefore the four lines OA, OB, OC, OD are all equal; and since they are perpendicular to the sides of the given square, *the circle described with O as centre, and OA as radius, will be inscribed in the square.*

PROP. IX.—PROBLEM.

About a given square (ABCD) to describe a circle.

Sol.—Draw the diagonals AC, BD intersecting in O (*see* diagram to Proposition VI.). *O is the centre of the required circle.*

Dem.—Since ABC is an isosceles triangle, and the angle B is right, each of the other angles is half a right angle; therefore BAO is half a right angle. In like manner ABO is half a right angle; hence the angle BAO equal ABO; therefore [I. VI.] AO is equal to OB. In like manner OB is equal to OC, and OC to OD. *Hence the circle described, with O as centre and OA as radius, will pass through the points B, C, D, and be described about the square.*

PROP. X.—PROBLEM.

To construct an isosceles triangle having each base angle double the vertical angle.

Sol.—Take any line AB. Divide it in C, so that the rectangle $AB \cdot BC$ shall be equal to AC^2 [II. XI.]. With A as centre, and AB as radius, describe the circle BDE, and in it place the chord BD equal to AC [I.]. Join AD. ADB *is a triangle fulfilling the required conditions.*

Dem.—Join CD. About the triangle ACD describe the circle CDE [V.]. Then, because the rectangle $AB \cdot BC$ is equal to AC^2 (const.), and that AC is equal to BD (const.); therefore the rectangle $AB \cdot BC$ is equal to BD^2. Hence [III. XXXII.] BD touches the circle ACD. Hence the angle BDC is equal to the angle A in the alternate segment [III. XXXII.]. To each add CDA, and we have the angle BDA equal to the sum of the angles CDA and A; but the exterior angle BCD of the triangle ACD is equal to the sum of the angles CDA and A. Hence the angle BDA is equal to BCD; but since AB is equal to AD, the angle BDA is equal to ABD; therefore the angle CBD is equal to BCD. Hence [I. VI.] BD is equal to CD; but BD is equal to AC (const.); therefore AC is equal to CD, and therefore [I. V.] the angle CDA is equal to A; but BDA has been proved to be equal to the sum of CDA and A. Hence BDA is double of A. *Hence each of the base angles of the triangle ABD is double of the vertical angle.*

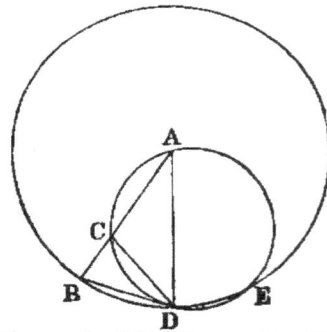

Exercises.

1. Prove that ACD is an isosceles triangle whose vertical angle is equal to three times each of the base angles.

2. Prove that BD is the side of a regular decagon inscribed in the circle BDE.

3. If DB, DE, EF be consecutive sides of a regular decagon inscribed in a circle, prove $BF - BD =$ radius of circle.

4. If E be the second point of intersection of the circle ACD with BDE, DE is equal to DB; and if AE, BE, CE, DE be joined, each of the triangles ACE, ADE is congruent with ABD.

5. AC is the side of a regular pentagon inscribed in the circle ACD, and EB the side of a regular pentagon inscribed in the circle BDE.

6. Since ACE is an isosceles triangle, $EB^2 - EA^2 = AB \cdot BC$—that is $= BD^2$; therefore $EB^2 - BD^2 = EA^2$—that is, *the square of the side of a pentagon inscribed in a circle exceeds the square of the side of the decagon inscribed in the same circle by the square of the radius.*

PROP. XI.—Problem.

To inscribe a regular pentagon in a given circle ($ABCDE$).

Sol.—Construct an isosceles triangle [x.], having each base angle double the vertical angle, and inscribe in the given circle a triangle ABD equiangular to it. Bisect the angles DAB, ABD by the lines AC, BE. Join EA, ED, DC, CB; *then the figure $ABCDE$ is a regular pentagon.*

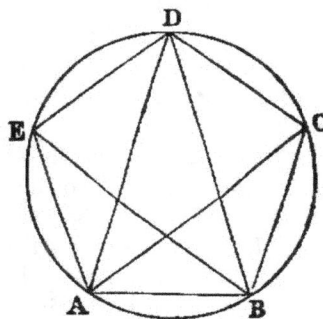

Dem.—Because each of the base angles BAD, ABD is double of the angle ADB, and the lines AC, BE bisect them, the five angles BAC, CAD, ADB, DBE, EBA are all equal; therefore the arcs on which they stand are equal; and therefore the five chords, AB, BC, CD, DE, EA are equal. Hence the figure $ABCDE$ is equilateral.

Again, because the arcs AB, DE are equal, adding the arc BCD to both, the arc $ABCD$ is equal to the arc $BCDE$, and therefore [III. xxvii.] the angles AED, BAE, which stand on them, are equal. In the same manner it can be proved that all the angles are equal; therefore the figure $ABCDE$ is equiangular. *Hence it is a regular pentagon.*

Exercises.

1. The figure formed by the five diagonals of a regular pentagon is another regular pentagon.

2. If the alternate sides of a regular pentagon be produced to meet, the five points of meeting form another regular pentagon.

3. Every two consecutive diagonals of a regular pentagon divide each other in extreme and mean ratio.

4. Being given a side of a regular pentagon, construct it.

5. Divide a right angle into five equal parts.

109

PROP. XII.—PROBLEM.

To describe a regular pentagon about a given circle (ABCDE).

Sol.—Let the five points A, B, C, D, E on the circle be the vertices of any inscribed regular pentagon: at these points draw tangents FG, GH, HI, IJ, JF: *the figure FGHIJ is a circumscribed regular pentagon.*

Dem.—Let O be the centre of the circle. Join OE, OA, OB. Now, because the angles A, E of the quadrilateral $AOEF$ are right angles [III. XVIII.], the sum of the two remaining angles AOE, AFE is two right angles. In like manner the sum of the angles AOB, AGB is two right angles; therefore the sum of AOE, AFE is equal to the sum of AOB, AGB; but the angles AOE, AOB are equal, because they stand on equal arcs AE, AB [III. XXVII.]. Hence the angle AFE is equal to AGB. In like manner the remaining angles of the figure $FGHIJ$ are equal. *Therefore it is equiangular.*

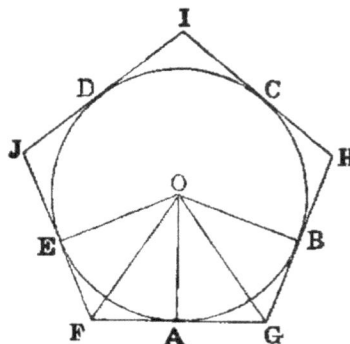

Again, join OF, OG. Now the triangles EOF, AOF have the sides AF, FE equal [III. XVII., Ex. 1], and FO common, and the base AO equal to the base EO. Hence the angle AFO is equal to EFO [I. VIII.]. Therefore the angle AFO is half the angle AFE. In like manner AGO is half the angle AGB; but AFE has been proved equal to AGB; hence AFO is equal to AGO, and FAO is equal to GAO, each being right, and AO common to the two triangles FAO, GAO; hence [I. XXVI.] the side AF is equal to AG; therefore GF is double AF. In like manner JF is double EF; but AF is equal to EF; hence GF is equal to JF. In like manner the remaining sides are equal; therefore the figure $FGHIJ$ is equilateral, and it has been proved equiangular. *Hence it is a regular pentagon.*

This Proposition is a particular case of the following general theorem, of which the proof is the same as the foregoing:—

"If tangents be drawn to a circle, at the angular points of an inscribed polygon of any number of sides, they will form a regular polygon of the same number of sides circumscribed to the circle."

PROP. XIII.—PROBLEM.

To inscribe a circle in a regular pentagon (ABCDE).

Sol.—Bisect two adjacent angles A, B by the lines AO, BO; *then O, the point of intersection of the bisectors, is the centre of the required circle.*

Dem.—Join CO, and let fall perpendiculars from O on the five sides of the pentagon. Now the triangles ABO, CBO have the side AB equal to BC (hyp.), and BO common, and the angle ABO equal to CBO (const.). Hence the angle BAO is equal to BCO [I. IV.]; but BAO is half BAE (const.). Therefore BCO is half BCD, and therefore CO bisects the angle BCD. In like manner it may be proved that DO bisects the angle D, and EO the angle E.

110

Again, the triangles BOF, BOG have the angle F equal to G, each being right; and OBF equal to OBG, because OB bisects the angle ABC (const.), and OB common; hence [I. XXVI.] OF is equal to OG. In like manner all the perpendiculars from O on the sides of the pentagon are equal; hence the circle whose centre is O, and radius OF, will touch all the sides of the pentagon, *and will therefore be inscribed in it.*

In the same manner a circle may be inscribed in any regular polygon.

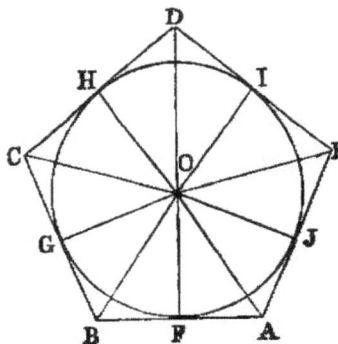

PROP. XIV.—PROBLEM.

To describe a circle about a regular pentagon (ABCDE).

Sol.—*Bisect two adjacent angles A, B by the lines AO, BO. Then O, the point of intersection of the bisectors, is the centre of the required circle.*

Dem.—Join OC, OD, OE. Then the triangles ABO, CBO have the side AB equal to BC (hyp.), BO common, and the angle ABO equal to CBO (const.). Hence the angle BAO is equal to BCO [I. IV.]; but the angle BAE is equal to BCD (hyp.); and since BAO is half BAE (const.), BCO is half BCD. Hence CO bisects the angle BCD. In like manner it may be proved that DO bisects CDE, and EO the angle DEA. Again, because the angle EAB is equal to ABC, their halves are equal. Hence OAB is equal to OBA; therefore [I. VI.] OA is equal to OB. In like manner the lines OC, OD, OE are equal to one another and to OA. Therefore the circle described with O as centre, and OA as radius, will pass through the points B, C, D, E, *and be described about the pentagon.*

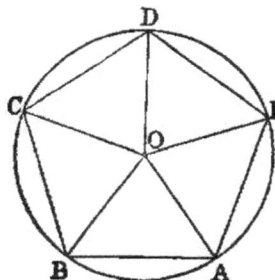

In the same manner a circle may be described about any regular polygon.

Propositions XIII., XIV. are particular cases of the following theorem:—

"A regular polygon of any number of sides has one circle inscribed in it, and another described about it, and both circles are concentric."

PROP. XV.—PROBLEM.

In a given circle (ABCDEF) to inscribe a regular hexagon.

Sol.—Take any point A in the circumference, and join it to O, the centre of the given circle; then with A as centre, and AO as radius, describe the circle OBF, intersecting the given circle in the points B, F. Join OB, OF, and produce AO, BO, FO to meet the given circle again in the points D, E, C. Join AB, BC, CD, DE, EF, FA; $ABCDEF$ is the required hexagon.

111

Dem.—Each of the triangles AOB, AOF is equilateral (*see* Dem., I. I.). Hence the angles AOB, AOF are each one-third of two right angles; therefore EOF is one-third of two right angles. Again, the angles BOC, COD, DOE are [I. XV.] respectively equal to the angles EOF, FOA, AOB. Therefore the six angles at the centre are equal, because each is one-third of two right angles. Therefore the six chords are equal [III. XXIX.]. *Hence the hexagon is equilateral.*

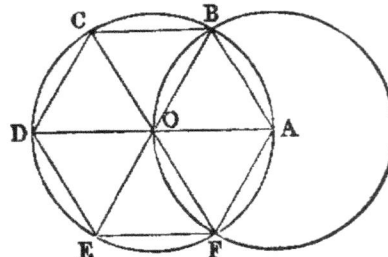

Again, since the arc AF is equal to ED, to each add the arc $ABCD$; then the whole arc $FABCD$ is equal to $ABCDE$; therefore the angles DEF, EFA which stand on these arcs are equal [III. XXVII.]. In the same manner it may be shown that the other angles of the hexagon are equal. Hence it is equiangular, *and is therefore a regular hexagon inscribed in the circle.*

Cor. 1.—The side of a regular hexagon inscribed in a circle is equal to the radius.

Cor. 2.—If three alternate angles of a hexagon be joined, they form an inscribed equilateral triangle.

Exercises.

1. The area of a regular hexagon inscribed in a circle is equal to twice the area of an equilateral triangle inscribed in the circle; and the square of the side of the triangle is three times the square of the side of the hexagon.

2. If the diameter of a circle be produced to C until the produced part is equal to the radius, the two tangents from C and their chord of contact form an equilateral triangle.

3. The area of a regular hexagon inscribed in a circle is half the area of an equilateral triangle, and three-fourths of the area of a regular hexagon circumscribed to the circle.

PROP. XVI.—PROBLEM.

To inscribe a regular polygon of fifteen sides in a given circle.

Sol.—Inscribe a regular pentagon $ABCDE$ in the circle [XI.], and also an equilateral triangle AGH [II.]. Join CG. *CG is a side of the required polygon.*

Dem.—Since $ABCDE$ is a regular pentagon, the arc ABC is $\frac{2}{5}$ths of the circumference; and since AGH is an equilateral triangle, the arc ABG is $\frac{1}{3}$rd of the circumference. Hence the arc GC, which is the difference between these two arcs, is equal to $\frac{2}{5}$ths $- \frac{1}{3}$rd, or $\frac{1}{15}$th of the entire circumference; *and therefore, if chords equal to GC [I.] be placed round the circle, we shall have a regular polygon of fifteen sides, or quindecagon, inscribed in it.*

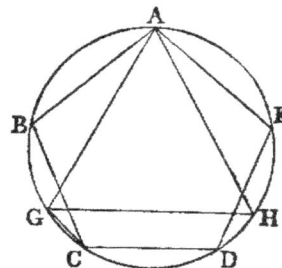

Scholium.—Until the year 1801 no regular polygon could be described by constructions employing the line and circle only, except those discussed in this Book, and those obtained from them by the continued bisection of the arcs of which their sides are the chords; but in

112

that year the celebrated Gauss proved that if $2^n + 1$ be a prime number, regular polygons of $2^n + 1$ sides are inscriptable by elementary geometry. For the case $n = 4$, which is the only figure of this class except the pentagon for which a construction has been given, see Note at the end of this work.

Questions for Examination on Book IV.

1. What is the subject-matter of Book IV.?
2. When is one rectilineal figure said to be inscribed in another?
3. When circumscribed?
4. When is a circle said to be inscribed in a rectilineal figure?
5. When circumscribed about it?
6. What is meant by reciprocal propositions? *Ans.* In reciprocal propositions, to every line in one there corresponds a point in the other; and, conversely, to every point in one there corresponds a line in the other.
7. Give instances of reciprocal propositions in Book IV.
8. What is a regular polygon?
9. What figures can be inscribed in, and circumscribed about, a circle by means of Book IV.?
10. What regular polygons has Gauss proved to be inscriptable by the line and circle?
11. What is meant by escribed circles?
12. How many circles can be described to touch three lines forming a triangle?
13. What is the centroid of a triangle?
14. What is the orthocentre?
15. What is the circumcentre?
16. What is the polar circle?
17. When is the polar circle imaginary?
18. What is the "nine-points circle"?
19. Why is it so called?
20. Name the special nine points through which it passes.
21. What three regular figures can be used in filling up the space round a point? *Ans.* Equilateral triangles, squares, and hexagons.
22. If the sides of a triangle be 13, 14, 15, what are the values of the radii of its inscribed and escribed circles?
23. What is the radius of the circumscribed circle?
24. What is the radius of its nine-points circle?
25. What is the distance between the centres of its inscribed and circumscribed circles?
26. If r be the radius of a circle, what is the area of its inscribed equilateral triangle?—of its inscribed square?—its inscribed pentagon?—its inscribed hexagon?—its inscribed octagon?—its inscribed decagon?
27. With the same hypothesis, find the sides of the same regular figures.

Exercises on Book IV.

1. If a circumscribed polygon be regular, the corresponding inscribed polygon is also regular, and conversely.
2. If a circumscribed triangle be isosceles, the corresponding inscribed triangle is isosceles, and conversely.
3. If the two isosceles triangles in Ex. 2 have equal vertical angles, they are both equilateral.
4. Divide an angle of an equilateral triangle into five equal parts.
5. Inscribe a circle in a sector of a given circle.
6. The line DE is parallel to the base BC of the triangle ABC: prove that the circles described about the triangles ABC, ADE touch at A.
7. The diagonals of a cyclic quadrilateral intersect in E: prove that the tangent at E to the circle about the triangle ABE is parallel to CD.
8. Inscribe a regular octagon in a given square.

113

9. A line of given length slides between two given lines: find the locus of the intersection of perpendiculars from its extremities to the given lines.

10. If the perpendicular to any side of a triangle at its middle point meet the internal and external bisectors of the opposite angle in the points D and E; prove that D, E are points on the circumscribed circle.

11. Through a given point P draw a chord of a circle so that the intercept EF may subtend a given angle X.

12. In a given circle inscribe a triangle having two sides passing through two given points, and the third parallel to a given line.

13. Given four points, no three of which are collinear; describe a circle which shall be equidistant from them.

14. In a given circle inscribe a triangle whose three sides shall pass through three given points.

15. Construct a triangle, being given—

1. The radius of the inscribed circle, the vertical angle, and the perpendicular from the vertical angle on the base.

2. The base, the sum or difference of the other sides, and the radius of the inscribed circle, or of one of the escribed circles.

3. The centres of the escribed circles.

16. If F be the middle point of the base of a triangle, DE the diameter of the circumscribed circle which passes through F, and L the point where a parallel to the base through the vertex meets DE: prove $DL \, . \, FE$ is equal to the square of half the sum, and $DF \, . \, LE$ equal to the square of half the difference of the two remaining sides.

17. If from any point within a regular polygon of n sides perpendiculars be let fall on the sides, their sum is equal to n times the radius of the inscribed circle.

18. The sum of the perpendiculars let fall from the angular points of a regular polygon of n sides on any line is equal to n times the perpendicular from the centre of the polygon on the same line.

19. If R denotes the radius of the circle circumscribed about a triangle ABC, r, r', r'', r''' the radii of its inscribed and escribed circles, δ, δ', δ'' the perpendiculars from its circumcentre on the sides; μ, μ', μ'' the segments of these perpendiculars between the sides and circumference of the circumscribed circle, we have the relations—

$$r' + r'' + r''' = 4R + r, \tag{1}$$

$$\mu + \mu' + \mu'' = 2R - r, \tag{2}$$

$$\delta + \delta' + \delta'' = R + r. \tag{3}$$

The relation (3) supposes that the circumcentre is inside the triangle.

20. Through a point D, taken on the side BC of a triangle ABC, is drawn a transversal EDF, and circles described about the triangles DBF, ECD. The locus of their second point of intersection is a circle.

21. In every quadrilateral circumscribed about a circle, the middle points of its diagonals and the centre of the circle are collinear.

22. Find on a given line a point P, the sum or difference of whose distances from two given points may be given.

23. Find a point such that, if perpendiculars be let fall from it on four given lines, their feet may be collinear.

24. The line joining the orthocentre of a triangle to any point P, in the circumference of its circumscribed circle, is bisected by the line of collinearity of perpendiculars from P on the sides of the triangle.

25. The orthocentres of the four triangles formed by any four lines are collinear.

26. If a semicircle and its diameter be touched by any circle, either internally or externally, twice the rectangle contained by the radius of the semicircle, and the radius of the tangential circle, is equal to the rectangle contained by the segments of any secant to the semicircle, through the point of contact of the diameter and touching circle.

114

27. If ρ, ρ' be the radii of two circles, touching each other at the centre of the inscribed circle of a triangle, and each touching the circumscribed circle, prove

$$\frac{1}{\rho} + \frac{1}{\rho'} = \frac{2}{r},$$

and state and prove corresponding theorems for the escribed circles.

28. If from any point in the circumference of the circle, circumscribed about a regular polygon of n sides, lines be drawn to its angular points, the sum of their squares is equal to $2n$ times the square of the radius.

29. In the same case, if the lines be drawn from any point in the circumference of the inscribed circle, prove that the sum of their squares is equal to n times the sum of the squares of the radii of the inscribed and the circumscribed circles.

30. State the corresponding theorem for the sum of the squares of the lines drawn from any point in the circumference of any concentric circle.

31. If from any point in the circumference of any concentric circle perpendiculars be let fall on all the sides of any regular polygon, the sum of their squares is constant.

32. For the inscribed circle, the constant is equal to $\dfrac{3n}{2}$ times the square of the radius.

33. For the circumscribed circle, the constant is equal to n times the square of the radius of the inscribed circle, together with $\frac{1}{2}n$ times the square of the radius of the circumscribed circle.

34. If the circumference of a circle whose radius is R be divided into seventeen equal parts, and AO be the diameter drawn from one of the points of division (A), and if $\rho_1, \rho_2 \ldots \ldots \rho_8$ denote the chords from O to the points of division, $A_1, A_2 \ldots \ldots A_8$ on one side of AO, then

$$\rho_1 \rho_2 \rho_4 \rho_8 = R^4; \text{ and } \rho_3 \rho_5 \rho_6 \rho_7 = R^4. \text{—Catalan.}$$

Dem.—Let the supplemental chords corresponding to ρ_1, ρ_2, &c., be denoted by r_1, r_2, &c.; then [III. xxxv. Ex. 2], we have

$$\rho_1 r_1 = R r_2,$$
$$\rho_2 r_2 = R r_4,$$
$$\rho_4 r_4 = R r_8,$$
$$\rho_8 r_8 = R r_1,$$

Hence $\qquad\qquad \rho_1 \rho_2 \rho_4 \rho_8 = R^4.$

And it may be proved in the same manner that

$$\rho_1 \rho_2 \rho_3 \rho_4 \rho_5 \rho_6 \rho_7 \rho_8 = R^8.$$

Therefore $\qquad\qquad \rho_3 \rho_5 \rho_6 \rho_7 = R^4.$

35. If from the middle point of the line joining any two of four concyclic points a perpendicular be let fall on the line joining the remaining two, the six perpendiculars thus obtained are concurrent.

36. The greater the number of sides of a regular polygon circumscribed about a given circle, the less will be its perimeter.

37. The area of any regular polygon of more than four sides circumscribed about a circle is less than the square of the diameter.

38. Four concyclic points taken three by three determine four triangles, the centres of whose nine-points circles are concyclic.

39. If two sides of a triangle be given in position, and if their included angle be equal to an angle of an equilateral triangle, the locus of the centre of its nine-points circle is a right line.

40. If, in the hypothesis and notation of Ex. 34, α, β denote any two suffixes whose sum is less than 8, and of which α is the greater,

$$\rho_\alpha \rho_\beta = R(\rho_{\alpha-\beta} + \rho_{\alpha+\beta}).$$

For instance, $\quad \rho_1 \rho_4 = R(\rho_3 + \rho_5)$ [III. xxxv., Ex. 7].

115

In the same case, if the suffixes be greater than 8,

$$\rho_\alpha \cdot \rho_\beta = R(\rho_{\alpha-\beta} - \rho_{17-\alpha-\beta}).$$

For instance, $\rho_8 \rho_2 = R(\rho_6 - \rho_7)$ [III. XXXV., Ex. 6].

41. Two lines are given in position: draw a transversal through a given point, forming with the given lines a triangle of given perimeter.

42. Given the vertical angle and perimeter of a triangle, construct it with either of the following data: 1. The bisector of the vertical angle; 2. the perpendicular from the vertical angle on the base; 3. the radius of the inscribed circle.

43. In a given circle inscribe a triangle so that two sides may pass through two given points, and that the third side may be a maximum or a minimum.

44. If s be the semiperimeter of a triangle, r', r'', r''', the radii of its escribed circles,

$$r'r'' + r''r''' + r'''r' = s^2.$$

45. The feet of the perpendiculars from the extremities of the base on either bisector of the vertical angle, the middle point of the base, and the foot of the perpendicular from the vertical angle on the base, are concyclic.

46. Given the base of a triangle and the vertical angle; find the locus of the centre of the circle passing through the centres of the escribed circles.

47. The perpendiculars from the centres of the escribed circles of a triangle on the corresponding sides are concurrent.

48. If AB be the diameter of a circle, and PQ any chord cutting AB in O, and if the lines AP, AQ intersect the perpendicular to AB at O, in D and E respectively, the points A, B, D, E are concyclic.

49. If the sides of a triangle be in arithmetical progression, and if R, r be the radii of the circumscribed and inscribed circles; then $6Rr$ is equal to the rectangle contained by the greatest and least sides.

50. Inscribe in a given circle a triangle having its three sides parallel to three given lines.

51. If the sides AB, BC, &c., of a regular pentagon be bisected in the points A', B', C', D', E', and if the two pairs of alternate sides, BC, AE; AB, DE, meet in the points A'', E'', respectively, prove

$$\triangle A''AE'' - \triangle A'AE' = \text{pentagon } A'B'C'D'E'.$$

52. In a circle, prove that an equilateral inscribed polygon is regular, and also an equilateral circumscribed polygon, if the number of sides be odd.

53. Prove also that an equiangular circumscribed polygon is regular, and an equiangular inscribed polygon, if the number of sides be odd.

54. The sum of the perpendiculars drawn to the sides of an equiangular polygon from any point inside the figure is constant.

55. Express the sides of a triangle in terms of the radii of its escribed circles.

116

BOOK V.

THEORY OF PROPORTION

DEFINITIONS.

Introduction.—Every proposition in the theories of ratio and proportion is true for all descriptions of magnitude. Hence it follows that the proper treatment is the Algebraic. It is, at all events, the easiest and the most satisfactory. Euclid's proofs of the propositions, in the *Theory of Proportion*, possess at present none but a historical interest, as no student reads them now. But although his demonstrations are abandoned, his propositions are quoted by every writer, and his nomenclature is universally adopted. For these reasons it appears to us that the best method is to state Euclid's definitions, explain them, or prove them when necessary, for some are theorems under the guise of definitions, and then supply simple algebraic proofs of his propositions.

I. A less magnitude is said to be a *part* or *submultiple* of a greater magnitude, when the less measures the greater—that is, when the less is contained a certain number of times exactly in the greater.

II. A greater magnitude is said to be a *multiple* of a less when the greater is measured by the less—that is, when the greater contains the less a certain number of times exactly.

III. Ratio is the mutual relation of two magnitudes of the same kind with respect to quantity.

IV. Magnitudes are said to have a *ratio* to one another when the less can be multiplied so as to exceed the greater.

These definitions require explanation, especially Def. III., which has the fault of conveying no precise meaning—being, in fact, unintelligible.

The following annotations will make them explicit:—

1. If an integer be divided into any number of equal parts, one, or the sum of any number of these parts, is called a *fraction*. Thus, if the line AB represent the integer, and if it be divided into four equal parts in the points C, D, E, then AC is $\frac{1}{4}$; AD, $\frac{2}{4}$; AE, $\frac{3}{4}$. Thus, a fraction is denoted by two numbers parted by a horizontal line; the lower, called the *denominator*, denotes the number of equal parts into which the integer is divided; and the upper, called the *numerator*, denotes the number of these equal parts which are taken. Hence it follows, that if the numerator be less than the denominator, the fraction is less than unity. If the numerator be equal to the denominator, the fraction is equal to unity; and if greater than the denominator, it is greater than unity. It is evident that a fraction is an abstract quantity—that is, that its value is independent of the nature of the integer which is divided.

2. If we divide each of the equal parts AC, CD, DE, EB into two equal parts, the whole, AB, will be divided into eight equal parts; and we see that $AC = \frac{2}{8}$; $AD = \frac{4}{8}$; $AE = \frac{6}{8}$; $AB = \frac{8}{8}$; Now, we saw in 1, that $AE = \frac{3}{4}$ of the integer, and we have just shown that it is equal to $\frac{6}{8}$. Hence $\frac{3}{4} = \frac{6}{8}$; but $\frac{6}{8}$ would be got from $\frac{3}{4}$ by multiplying its terms (numerator and denominator) by 2. Hence we infer generally that multiplying the terms of any fraction by 2 does not alter its value. In like manner it may be shown that multiplying the terms of

117

a fraction by any whole number does not alter its value. Hence it follows conversely, that dividing the terms of a fraction by a whole number does not alter the value. Hence we have the following important and fundamental theorem:—*Two transformations can be made on any fraction without changing its value; namely, its terms can be either multiplied or divided by any whole number, and in either case the value of the new fraction is equal to the value of the original one.*

3. If we take any number, such as 3, and multiply it by any whole number, the product is called a *multiple* of 3. Thus 6, 9, 12, 15, &c., are multiples of 3; but 10, 13, 17, &c., are not, because the multiplication of 3 by any whole number will not produce them. Conversely, 3 is a *submultiple*, or *measure*, or *part* of 6, 9, 12, 15, &c., because it is contained in each of these without a remainder; but not of 10, 13, 17, &c., because in each case it leaves a remainder.

4. If we consider two magnitudes of the same kind, such as two lines AB, CD, and if we suppose that AB is equal to $\frac{3}{4}$ of CD, it is evident, if AB be divided into 3 equal parts, and CD into 4 equal parts, that one of the parts into which AB is divided is equal to one of the parts into which

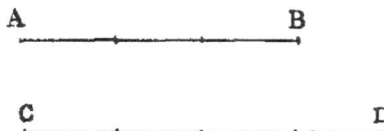

CD is divided. And as there are 3 parts in AB, and 4 in CD, we express this relation by saying that AB has to CD the ratio of 3 to 4; and we denote it thus, 3 : 4. Hence the ratio 3 : 4 expresses the same idea as the fraction $\frac{3}{4}$. In fact, both are different ways of expressing and writing the same thing. When written 3 : 4 it is called a ratio, and when $\frac{3}{4}$ a fraction. *In the same manner it can be shown that every ratio whose terms are commensurable can be converted into a fraction; and, conversely, every fraction can be turned into a ratio.*

From this explanation we see that the ratio of any two commensurable magnitudes is the same as the ratio of the numerical quantities which denote these magnitudes. Thus, the ratio of two commensurable lines is the ratio of the numbers which express their lengths, measured with the same unit. And this may be extended to the case where the lines are incommensurable. Thus, if a be the side and b the diagonal of a square, the ratio of $a : b$ is

$$\frac{a}{b}, \text{ or } \frac{1}{\sqrt{2}}.$$

When two quantities are incommensurable, such as the diagonal and the side of a square, although their ratio is not equal to that of any two commensurable numbers, yet a series of pairs of fractions can be found whose difference is continually diminishing, and which ultimately becomes indefinitely small; such that the ratio of the incommensurable quantities is greater than one, and less than the other fraction of each pair. These fractions are called convergents. By their means we can approximate as nearly as we please to the exact value of the ratio. In the case of the diagonal and the side of a square, the following are the pairs of convergents:—

$$\frac{14}{10}, \frac{15}{10}; \quad \frac{141}{100}, \frac{142}{100}; \quad \frac{1414}{1000}, \frac{1415}{1000}; \quad \&c.,$$

and the ratio is intermediate to each pair. It is evident we may continue the series as far as we please. Now if we denote the first of any of the foregoing pairs of fractions by $\frac{m}{n}$, the second will be $\frac{m+1}{n}$; and in general, in the case of two incommensurable quantities, two fractions $\frac{m}{n}$ and $\frac{m+1}{n}$ can always be found, where n can be made as large as we please, one of which is less and the other greater than the true value of the ratio. For let a and b be the incommensurable quantities; then, evidently, we cannot find two multiples na, mb, such that $na = mb$. In this case, take any multiple of a, such as na, then this quantity must lie between some two consecutive multiples of b, such as mb, and $(m+1)b$; therefore $\frac{na}{mb}$ is greater than unity, and $\frac{na}{(m+1)b}$ less than unity. Hence $\frac{a}{b}$ lies between $\frac{m}{n}$ and $\frac{m+1}{n}$. Now, since the difference between $\frac{m}{n}$ and $\frac{m+1}{n}$ namely, $\frac{1}{n}$ becomes small as n increases, we see that the difference between the ratio of two incommensurable quantities and that of two commensurable numbers

m and n can be made as small as we please. Hence, ultimately, the ratio of incommensurable quantities may be regarded as the limit of the ratio of commensurable quantities.

5. The two terms of a ratio are called the *antecedent* and the *consequent*. These correspond to the numerator and the denominator of a fraction. Hence we have the following definition:— "*A ratio is the fraction got by making the antecedent the numerator and the consequent the denominator.*"

6. The *reciprocal* of a ratio is the ratio obtained by interchanging the antecedent and the consequent. Thus, 4 : 3 is the reciprocal of the ratio 3 : 4. Hence we have the following theorem:—"*The product of a ratio and its reciprocal is unity.*"

7. If we multiply any two numbers, as 5 and 7, by any number such as 4, the products 20, 28 are called *equimultiples* of 5 and 7. In like manner, 10 and 15 are equimultiples of 2 and 3, and 18 and 30 of 3 and 5, &c.

V. The first of four magnitudes has to the second the same ratio which the third has to the fourth, when any equimultiples whatsoever of the first and third being taken, and any equimultiples whatsoever of the second and fourth, if, according as the multiple of the first is greater than, equal to, or less than the multiple of the second, the multiple of the third is greater than, equal to, or less than the multiple of the fourth.

VI. Magnitudes which have the same ratio are called proportionals. When four magnitudes are proportionals, it is usually expressed by saying, "The first is to the second as the third is to the fourth."

VIII. Analogy or proportion is the similitude of ratios.

We have given the foregoing definitions in the order of Euclid, as given by Simson, Lardner, and others;[2] but it is evidently an inverted order; for VI. VIII. are definitions of proportion, and V. is only a test of proportion, and is not a definition but a theorem, and one which, instead of being taken for granted, requires proof. The following explanations will give the student clear conceptions of their meaning:—

1. If we take two ratios, such as 6 : 9 and 10 : 15, which are each equal to the same thing (in this example each is equal to $\frac{2}{3}$), they are equal to one another (I. Axiom I.). Then we may write it thus—

$$6 : 9 = 10 : 15.$$

This would be the most intelligible way, but it is not the usual one, which is as follows:— 6 : 9 :: 10 : 15. In this form it is called a *proportion*. Hence a proportion consists of two ratios which are asserted by it to be equal. Its four terms consist of two antecedents and two consequents. The 1st and 3rd terms are the *antecedents*, and the 2nd and 4th the *consequents*. Also the first and last terms are called the *extremes*, and the two middle terms the *means*.

2. Since a proportion consists of two equal ratios, and each ratio can be written as a fraction, whenever we have a proportion such as

$$a : b :: c : d,$$

we can write it in the form of two equal fractions. Thus:

$$\frac{a}{b} = \frac{c}{d}.$$

Conversely, an equation between two fractions can be put into a proportion. By means of these simple principles all the various properties of proportion can be proved in the most direct and easy manner.

3. If we take the proportion $a : b :: c : d$, and multiply the first and third terms, each by m, and second and fourth, each by n, we get the four multiples, ma, nb, mc, nd; and we want to prove that if ma is greater than nb, mc is greater than nd; if equal, equal; and if less, less.

Dem.—Since $\qquad\qquad a : b :: c : d,$

we have $\qquad\qquad\qquad \dfrac{a}{b} = \dfrac{c}{d}.$

_____

[2]Except that VIII. is put before VII., because it relates, as V. and VI., to the equality of ratios, whereas VII. is a test of their inequality.

Hence, multiplying each by $\dfrac{m}{n}$ we get

$$\frac{ma}{nb} = \frac{mc}{nd}.$$

Now, it is evident that if $\dfrac{ma}{nb}$ is greater than unity, $\dfrac{mc}{nd}$ is greater than unity; but if $\dfrac{ma}{nb}$ is greater than unity, ma is greater than nb; and if $\dfrac{mc}{nd}$ is greater than unity, mc is greater than nd. In like manner, if ma be equal to nb, mc is equal to nd; and if less, less.

The foregoing is an easy proof of the converse of the theorem which is contained in Euclid's celebrated Fifth Definition.

Next, to prove Euclid's theorem—that if, according as the multiple of the first of four magnitudes is greater than, equal to, or less than the multiple of the second, the multiple of the third is greater than, equal to, or less than the multiple of the fourth; the ratio of the first to the second is equal to the ratio of the third to the fourth.

Dem.—Let, a, b, c, d be the four magnitudes. First suppose that a and b are commensurable, then it is evident that we can take multiples na, mb, such that $na = mb$. Hence, by hypothesis, $nc = md$. Thus,

$$\frac{na}{mb} = 1, \; \frac{nc}{md} = 1;$$

therefore

$$\frac{a}{b} = \frac{c}{d}.$$

Next, suppose a and b are incommensurable. Then, as in a recent note, we can find two numbers m and n, such that $\dfrac{na}{mb}$ is greater than unity, but $\dfrac{na}{(m+1)b}$ less than unity. Hence $\dfrac{a}{b}$ lies between $\dfrac{m}{n}$ and $\dfrac{m+1}{n}$. Now, since by hypothesis, when $\dfrac{na}{mb}$ is greater than unity, $\dfrac{nc}{md}$ is greater than unity; and when $\dfrac{na}{(m+1)b}$ is less than unity, $\dfrac{nc}{(m+1)d}$ is less than unity. Hence, since $\dfrac{a}{b}$ lies between $\dfrac{m}{n}$ and $\dfrac{m+1}{n}$, $\dfrac{c}{d}$ lies between the same quantities. Therefore the difference between $\dfrac{a}{b}$ and $\dfrac{c}{d}$ is less than $\dfrac{1}{n}$; and since n may be as large as we please, the difference is nothing; therefore

$$\frac{a}{b} = \frac{c}{d}.$$

VII. When of the multiples of four magnitudes (taken as in Def. V.) the multiple of the first is greater than that of the second, but the multiple of the third not greater than that of the fourth, the first has to the second a greater ratio than the third has to the fourth.

This, instead of being a definition, is a theorem. We have altered the last clause from that given in Simson's Euclid, which runs thus:—"The first is *said* to have to the second a greater ratio than the third has to the fourth." This is misleading, as it implies that it is, by convention, that the first ratio is greater than the second, whereas, in fact, such is not the case; for it follows from the hypothesis that the first ratio is greater than the second; and if it did not, it could not be made so by definition. We have made a similar change in the enunciation of the Fifth Definition.

Let a, b, c, d be the four magnitudes, and m and n the multiples taken, it is required to prove, that if ma be greater than nb, but mc not greater than nd, that the ratio $a : b$ is greater than the ratio $c : d$.

Dem.—Since ma is greater than nb, but mc not greater than nd, it is evident that

$$\frac{ma}{nb} \text{ is greater than } \frac{mc}{nd};$$

therefore

$$\frac{a}{b} \text{ is greater than } \frac{c}{d};$$

that is, the ratio $a : b$ is greater than the ratio $c : d$.

120

IX. Proportion consists of three terms at least.

This has the same fault as some of the others—it is not a definition, but an inference. It occurs when the means in a proportion are equal, so that, in fact, there are four terms. As an illustration, let us take the numbers 4, 6, 9. Here the ratio of 4 : 6 is $\frac{2}{3}$, and the ratio of 6 : 9 is $\frac{2}{3}$, so that 4, 6, 9 are continued proportionals; but, in reality, there are four terms, for the full proportion is 4 : 6 :: 6 : 9.

X. When three magnitudes are continual proportionals, the first is said to have to the third the *duplicate* ratio of that which it has to the second.

XI. When four magnitudes are continual proportionals, the first is said to have to the fourth the *triplicate* ratio of that which it has to the second.

XII. When there is any number of magnitudes of the same kind greater than two, the first is said to have to the last the ratio compounded of the ratios of the first to the second, of the second to the third, of the third to the fourth, &c.

We have placed these definitions in a group; but their order is inverted, and, as we shall see, Def. XII. is a theorem, and X. and XI. are only inferences from it.

1. If we have two ratios, such as 5 : 7 and 3 : 4, and if we convert each ratio into a fraction, and multiply these fractions together, we get a result which is called the ratio compounded of the two ratios; viz. in this example it is $\frac{15}{28}$, or 15 : 28. It is evident we get the same result if we multiply the two antecedents together for a new antecedent, and the two consequents for a new consequent. Hence we have the following definition:—"*The ratio compounded of any number of ratios it the ratio of the product of all the antecedents to the product of all the consequents.*"

2. To prove the theorem contained in Def. XII.

Let the magnitudes be a, b, c, d. Then the ratio of

$$\text{1st : 2nd} = \frac{a}{b},$$
$$\text{2nd : 3rd} = \frac{b}{c},$$
$$\text{3rd : 4th} = \frac{c}{d}.$$

Hence the ratio compounded of the ratio of 1st : 2nd, of 2nd : 3rd, 3rd : 4th

$$= \frac{abc}{bcd} = \frac{a}{d} = \text{ratio of 1st : 4th.}$$

3. If three magnitudes be proportional, the ratio of the 1st : 3rd is equal to the square of the ratio of the 1st : 2nd. For the ratio of the 1st : 3rd is compounded of the ratio of the 1st : 2nd, and of the ratio of the 2nd : 3rd; and since these ratios are equal, the ratio compounded of them will be equal to the square of one of them.

Or thus: Let the proportionals be a, b, c, that is, let $a : b :: b : c$; hence we have

$$\frac{b}{c} = \frac{a}{b}.$$

And multiplying each by $\frac{a}{b}$, we get

$$\frac{a}{c} = \frac{a^2}{b^2};$$

or $a : c :: a^2 : b^2$—that is, 1st : 3rd :: square of 1st : square of 2nd. Now, the ratio of 1st : 3rd is, by Def. X., the duplicate ratio of 1st : 2nd. Hence the duplicate ratio of two magnitudes means the square of their ratio, or, what is the same thing, the ratio of their squares (see Book VI. XX.).

4. If four magnitudes be continual proportionals, the ratio of 1st : 4th is equal to the cube of the ratio of 1st : 2nd. This may be proved exactly like 3. Hence we see that what Euclid calls triplicate ratio of two magnitudes is the ratio of their cubes, or the cube of their ratio.

We also see that there is no necessity to introduce extraneous magnitudes for the purpose of defining duplicate and triplicate ratios, as Euclid does. In fact, the definitions by squares and cubes are more explicit.

XIII. In proportionals, the antecedent terms are called *homologous* to one another; as also the consequents to one another.

If one proportion be given, from it an indefinite number of other proportions can be inferred, and a great part of the theory of proportion consists in proving the truth of these derived proportions. Geometers make use of certain technical terms to denote the most important of these processes. We shall indicate these terms by including them in parentheses in connexion with the Propositions to which they refer. They are useful as indicating, by one word, the whole enunciation of a theorem.

Every Proposition in the Fifth Book is a Theorem.

PROP. I.—THEOREM.

If any number of magnitudes of the same kind (a, b, c, &c.), be equimultiples of as many others (a', b', c', &c.), then the sum of the first magnitudes ($a + b + c$, &c.) shall be the same multiple of the sum of the second which any magnitude of the first system is of the corresponding magnitude of the second system.

Dem.—Let m denote the multiple which the magnitudes of the first system are of those of the second system.

Then we have
$$a = ma' \text{ (hyp.)},$$
$$b = mb',$$
$$c = mc'.$$
$$\text{&c., &c.}$$

Hence, by addition,
$$(a + b + c + \text{&c.}) = m(a' + b' + c', \text{&c.}).$$

PROP. II.—THEOREM.

If two magnitudes of the same kind (a, b) be the same multiples of another (c) which two corresponding magnitudes (a', b') are of another (c'), then the sum of the two first is the same multiple of their submultiple which the sum of their corresponding magnitudes is of their submultiple.

Dem.—Let m and n be the multiples which a and b are of c.
Then we have
$$a = mc \text{ and } a' = mc',$$
$$b = nc \text{ and } b' = nc'.$$

Therefore
$$(a + b) = (m + n)c, \text{ and } (a' + b') = (m + n)c'.$$

Hence $a + b$ is the same multiple of c that $a' + b'$ is of c'.

This Proposition is evidently true for any number of multiples.

PROP. III.—THEOREM.

If two magnitudes (a, b) be equimultiples of two others (a′, b′); then any equimultiples of the first magnitudes (a, b) will be also equimultiples of the second magnitudes (a′, b′).

Dem.—Let m denote the multiples which a, b are of $a′$, $b′$; then we have

$$a = ma′, \quad b = mb′.$$

Hence, multiplying each equation by n, we get

$$na = mna′, \quad nb = mnb′.$$

Hence, na, nb are equimultiples of $a′$, $b′$.

PROP. IV.—THEOREM.

If four magnitudes be proportional, and if any equimultiples of the first and third be taken, and any other equimultiples of the second and fourth; then the multiple of the first : the multiple of the second :: the multiple of the third : the multiple of the fourth.

Let $a : b :: c : d$; then $ma : nb :: mc : nd$.

Dem.—We have $a : b :: c : d$ (hyp.);

therefore $\qquad\qquad \dfrac{a}{b} = \dfrac{c}{d}.$

Hence, multiplying each fraction by $\dfrac{m}{n}$, we get

$$\frac{ma}{nb} = \frac{mc}{nd};$$

therefore $\qquad\qquad ma : nb :: mc : nd.$

PROP. V.—THEOREM.

If two magnitudes of the same kind (a, b) be the same multiples of another (c) which two corresponding magnitudes (a′, b′) are of another (c′), then the difference of the two first is the same multiple of their submultiple (c), which the difference of their corresponding magnitudes is of their submultiple (c′) (compare Proposition II.).

Dem.—Let m and n be the multiples which a and b are of c.

Then we have $\qquad\qquad a = mc$, and $a′ = mc′$,

$$b = nc, \text{ and } b′ = nc′.$$

Therefore $(a - b) = (m - n)c$, and $(a′ - b′) = (m - n)c′$. *Hence $a - b$ is the same multiple of c that $a′ - b′$ is of $c′$.*

Cor.—If $a - b = c$, $a′ - b′ = c′$; for if $a - b = c$, $m - n = 1$.

123

PROP. VI.—THEOREM.

If a magnitude (a) be the same multiple of another (b), which a magnitude (a') taken from the first is of a magnitude (b') taken from the second, the remainder is the same multiple of the remainder that the whole is of the whole (compare Proposition I.).

Dem.—Let m denote the multiples which the magnitudes a, a' are of b, b'; then we have

$$a = mb,$$
$$a' = mb'.$$

Hence
$$(a - a') = m(b - b').$$

PROP. A.—THEOREM (SIMSON).

If two ratios be equal, then according as the antecedent of the first ratio is greater than, equal to, or less than its consequent, the antecedent of the second ratio is greater than, equal to, or less than its consequent.

Dem.—Let $a : b :: c : d$;

then
$$\frac{a}{b} = \frac{c}{d};$$

and if a be greater than b, $\frac{a}{b}$ is greater than unity; therefore $\frac{c}{d}$ is greater than unity, and c is greater than d.

In like manner, if a be equal to b, c is equal to d, and if less, less.

PROP. B.—THEOREM (SIMSON).

If two ratios are equal their reciprocals are equal (invertendo).

Let $\qquad a : b :: c : d$, then $b : a :: d : c$.

Dem.—Since $\qquad a : b :: c : d$;

then
$$\frac{a}{b} = \frac{c}{d};$$

therefore
$$1 \div \frac{a}{b} = 1 \div \frac{c}{d},$$

or
$$\frac{b}{a} = \frac{d}{c}$$

Hence $\qquad b : a :: d : c.$

Prop. C.—Theorem (Simson).

If the first of four magnitudes be the same multiple of the second which the third is of the fourth, the first is to the second as the third is to the fourth.

Let $a = mb$, $c = md$; then $a : b :: c : d$.

Dem.—Since $a = mb$, we have $\dfrac{a}{b} = m$.

In like manner, $\dfrac{c}{d} = m$; therefore $\dfrac{a}{b} = \dfrac{c}{d}$.

Hence $a : b :: c : d$.

Prop. D.—Theorem (Simson).

If the first be to the second as to the third is to the fourth, and if the first be a multiple or submultiple of the second, the third is the same multiple or submultiple of the fourth.

1. Let $a : b :: c : d$, and let a be a multiple of b, then c is the same multiple of d.

Dem.—Let $a = mb$, then $\dfrac{a}{b} = m$;

but $\dfrac{a}{b} = \dfrac{c}{d}$; therefore $\dfrac{c}{d} = m$, and $c = md$.

2. Let $a = \dfrac{b}{n}$, then $\dfrac{a}{b} = \dfrac{1}{n}$;

therefore $\dfrac{c}{d} = \dfrac{1}{n}$,

Hence $c = \dfrac{d}{n}$.

PROP. VII.—Theorem.

1. Equal magnitudes have equal ratios to the same magnitude.
2. The same magnitude has equal ratios to equal magnitudes.

Let a and b be equal magnitudes, and c any other magnitude.

Then 1. $a : c :: b : c$,

2. $c : a :: c : b$.

Dem.—Since $a = b$, dividing each by c, we have

$$\frac{a}{c} = \frac{b}{c};$$

therefore $a : c :: b : c$.

125

Again, since $a = b$, dividing c by each, we have

$$\frac{c}{a} = \frac{c}{b};$$

therefore $\qquad\qquad\qquad\qquad\qquad c : a :: c : b.$

Observation.—2 follows at once from 1 by Proposition B.

PROP. VIII.—THEOREM.

1. *Of two unequal magnitudes, the greater has a greater ratio to any third magnitude than the less has;* 2. *any third magnitude has a greater ratio to the less of two unequal magnitudes than it has to the greater.*

1. Let a be greater than b, and let c be any other magnitude of the same kind, then the ratio $a : c$ is greater than the ratio $b : c$.

Dem.—Since a is greater than b, dividing each by c,

$$\frac{a}{c} \text{ is greater than } \frac{b}{c};$$

therefore the ratio $a : c$ is greater than the ratio $b : c$.

2. To prove that the ratio $c : b$ is greater than the ratio $c : a$.

Dem.—Since b is less than a, the quotient which is the result of dividing any magnitude by b is greater than the quotient which is got by dividing the same magnitude by a;

therefore $\qquad\qquad\qquad \frac{c}{b} \text{ is greater than } \frac{c}{a}.$

Hence the ratio $c : b$ is greater than the ratio $c : a$.

PROP. IX.—THEOREM.

Magnitudes which have equal ratios to the same magnitude are equal to one another; 2. *magnitudes to which the same magnitude has equal ratios are equal to one another.*

1. If $a : c :: b : c$, to prove $a = b$.

Dem.—Since $\qquad\qquad\qquad a : c :: b : c,$

$$\frac{a}{c} = \frac{b}{c}.$$

Hence, multiplying each by c, we get $a = b$.

2. If $c : a :: c : b$, to prove $a = b$.

Dem.—Since $\qquad\qquad\qquad c : a :: c : b,$

by inversion, $\qquad\qquad\qquad a : c :: b : c;$

therefore $\qquad\qquad\qquad\qquad a = b. \qquad$ [1].

126

PROP. X.—Theorem.

Of two unequal magnitudes, that which has the greater ratio to any third is the greater of the two; and that to which any third has the greater ratio is the less of the two.

1. If the ratio $a : c$ be greater than the ratio $b : c$, to prove a greater than b.

Dem.—Since the ratio $a : c$ is greater than the ratio $b : c$,

$$\frac{a}{c} \text{ is greater than } \frac{b}{c}.$$

Hence, multiplying each by c, we get a greater than b.

2. If the ratio $c : b$ is greater than the ratio $c : a$, to prove b is less than a.

Dem.—Since the ratio $c : b$ is greater than the ratio $c : a$,

$$\frac{c}{b} \text{ is greater than } \frac{c}{a}.$$

Hence $\qquad\qquad 1 \div \dfrac{c}{b} \text{ is less than } 1 \div \dfrac{c}{a},$

that is, $\qquad\qquad \dfrac{b}{c} \text{ is less than } \dfrac{a}{c}.$

Hence, multiplying each by c, we get

$$b \text{ less than } a.$$

PROP. XI.—Theorem.

Ratios that are equal to the same ratio are equal to one another.

Let $a : b :: e : f$, and $c : d :: e : f$, to prove $a : b :: c : d$.

Dem.—Since $\qquad\qquad a : b :: e : f$,

$$\frac{a}{b} = \frac{e}{f}.$$

In like manner, $\qquad\qquad \dfrac{c}{d} = \dfrac{e}{f}.$

Hence $\qquad\qquad \dfrac{a}{b} = \dfrac{c}{d} \qquad$ [I., Axiom I.],

and $\qquad\qquad a : b :: c : d.$

127

PROP. XII.—THEOREM.

If any number of ratios be equal to one another, any one of these equal ratios is equal to the ratio of the sum of all the antecedents to the sum of all the consequents.

Let the ratios $a : b$, $c : d$, $e : f$, be all equal to one another; it is required to prove that any of these ratios is equal to the ratio $a + c + e : b + d + f$.

Dem.—By hypotheses,
$$\frac{a}{b} = \frac{c}{d} = \frac{e}{f}.$$

Since these fractions are all equal, let their common value be r; then we have
$$\frac{a}{b} = r, \quad \frac{c}{d} = r, \quad \frac{e}{f} = r;$$

therefore
$$a = br,$$
$$c = dr,$$
$$e = fr;$$

therefore
$$a + c + e = (b + d + f)r.$$

Hence
$$\frac{a + c + e}{b + d + f} = r;$$

therefore
$$\frac{a}{b} = \frac{a + c + e}{b + d + f},$$

and
$$a : b :: a + c + e : b + d + f.$$

Cor.—With the same hypotheses, if l, m, n be any three multipliers, $a : b :: la + mc + ne : lb + md + nf$.

PROP. XIII.—THEOREM.

If two ratios are equal, and if one of them be greater than any third ratio, then the other is also greater than that third ratio.

If $a : b :: c : d$, but the ratio of $c : d$ greater than the ratio of $e : f$; then the ratio of $a : b$ is greater than the ratio of $e : f$.

Dem.—Since the ratio of $c : d$ is greater than the ratio of $e : f$,
$$\frac{c}{d} \text{ is greater than } \frac{e}{f}.$$

Again, since
$$a : b :: c : d,$$
$$\frac{a}{b} = \frac{c}{d};$$

therefore
$$\frac{a}{b} \text{ is greater than } \frac{e}{f}.$$

or the ratio of $a : b$ is greater than the ratio of $e : f$.

PROP. XIV.—Theorem.

If two ratios be equal, then, according as the antecedent of the first ratio is greater than, equal to, or less than the antecedent of the second, the consequent of the first is greater than, equal to, or less than the consequent of the second.

Let $a : b :: c : d$; then if a be greater than c, b is greater than d; if equal, equal; if less, less.

Dem.—Since
$$a : b :: c : d.$$

we have
$$\frac{a}{b} = \frac{c}{d},$$

and multiplying each by $\dfrac{b}{c}$ we get

$$\frac{a}{b} \times \frac{b}{c} = \frac{c}{d} \times \frac{b}{c},$$

or
$$\frac{a}{c} = \frac{b}{d};$$

therefore
$$a : c :: b : d.$$

Hence, Proposition [A], if a be greater than c, b is greater than d; if equal, equal; and if less, less.

PROP. XV.—Theorem.

Magnitudes have the same ratio which all equimultiples of them have.

Let a, b be two magnitudes, then the ratio $a : b$ is equal to the ratio $ma : mb$.
Dem.—The ratio $a : b = \dfrac{a}{b}$, and the ratio of $ma : mb = \dfrac{ma}{mb}$; but since the value of a fraction is not altered by multiplying its numerator and denominator by the same number,

$$\frac{a}{b} = \frac{ma}{mb};$$

therefore
$$a : b :: ma : mb.$$

PROP. XVI—Theorem.

If four magnitudes of the same kind be proportionals they are also proportionals by alternation (alternando).

Let
$$a : b :: c : d, \text{ then } a : c :: b : d.$$
Dem.—Since
$$a : b :: c : d,$$

$$\frac{a}{b} = \frac{c}{d},$$

129

and multiplying each by $\dfrac{b}{c}$, we get

$$\frac{a}{b} \cdot \frac{b}{c} = \frac{c}{d} \cdot \frac{b}{c},$$

or

$$\frac{a}{c} = \frac{b}{d};$$

therefore

$$a : c :: b : d.$$

PROP. XVII.—Theorem.

If four magnitudes be proportional, the difference between the first and second : the second :: the difference between the third and fourth : the fourth (dividendo).

Let

$$a : b :: c : d : \text{ then } a - b : b :: c - d : d;$$

Dem.—Since

$$a : b :: c : d,$$

$$\frac{a}{b} = \frac{c}{d};$$

therefore

$$\frac{a}{b} - 1 = \frac{c}{d} - 1,$$

or

$$\frac{a - b}{b} = \frac{c - d}{d};$$

therefore

$$a - b : b :: c - d : d.$$

PROP. XVIII.—Theorem.

If four magnitudes be proportionals, the sum of the first and second : the second :: the sum of the third and fourth : the fourth (componendo).

Let

$$a : b :: c : d; \text{ then } a + b : b :: c + d : d.$$

Dem.—Since

$$a : b :: c : d,$$

$$\frac{a}{b} = \frac{c}{d};$$

therefore

$$\frac{a}{b} + 1 = \frac{c}{d} + 1,$$

or

$$\frac{a + b}{b} = \frac{c + d}{d};$$

therefore

$$a + b : b :: c + d : d.$$

PROP. XIX.—THEOREM.

If a whole magnitude be to another whole at a magnitude taken from the first it to a magnitude taken from the second, the first remainder : the second remainder :: the first whole : the second whole.

Let $a : b :: c : d$, c and d being less than a and b;

then $\qquad a - c : b - d :: a : b.$

Dem.—Since $\qquad a : b :: c : d,$

then $\qquad a : c :: b : d$ [*alternando*],

and $\qquad c : a :: d : b$ [*invertendo*];

therefore $\qquad \dfrac{c}{a} = \dfrac{d}{b},$

and $\qquad 1 - \dfrac{c}{a} = 1 - \dfrac{d}{b},$

or $\qquad \dfrac{a - c}{a} = \dfrac{b - d}{b}.$

Hence $\qquad a - c : b - d :: a : b.$

PROP. E.—THEOREM (SIMSON).

If four magnitudes be proportional, the first : its excess above the second :: the third : its excess above the fourth (convertendo).

Let $\quad a : b :: c : d$; then $a : a - b :: c : c - d.$

Dem.—Since $\qquad a : b :: c : d,$

$$\frac{a}{b} = \frac{c}{d};$$

therefore $\qquad \dfrac{a - b}{b} = \dfrac{c - d}{d} \qquad$ [Dem. of XVII.],

therefore $\qquad \dfrac{a}{b} \div \dfrac{a - b}{b} = \dfrac{c}{d} \div \dfrac{c - d}{d},$

or $\qquad \dfrac{a}{a - b} = \dfrac{c}{c - d},$

therefore $\qquad a : a - b :: c : c - d.$

PROP. XX.—THEOREM.

If there be two sets of three magnitudes, which taken two by two in direct order have equal ratios, then if the first of either set be greater than the third, the first of the other set is greater than the third; if equal, equal; and if less, less.

Let a, b, c; a', b', c' be the two sets of magnitudes, and let the ratio $a : b = a' : b'$, and $b : c = b' : c'$; then, if a be greater than, equal to, or less than c, a' will be greater than, equal to, or less than c'.

Dem.—Since
$$a : b :: a' : b',$$

we have
$$\frac{a}{b} = \frac{a'}{b'},$$

In like manner,
$$\frac{b}{c} = \frac{b'}{c'},$$

Hence
$$\frac{a}{b} \times \frac{b}{c} = \frac{a'}{b'} \times \frac{b'}{c'},$$

or
$$\frac{a}{c} = \frac{a'}{c'}.$$

Therefore if a be greater than c, a' is greater than c'; if equal, equal; and if less, less.

PROP. XXI.—Theorem.

If there be two sets of three magnitudes, which taken two by two in transverse order have equal ratios; then, if the first of either set be greater than the third, the first of the other set is greater than the third; if equal, equal; and if less, less.

Let a, b, c; a', b', c' be the two sets of magnitudes, and let the ratio $a : b = b' : c'$, and $b : c = a' : b'$. Then, if a be greater than, equal to, or less than c, a' will be greater than, equal to, or less than c'.

Dem.—Since
$$a : b :: b' : c',$$

we have
$$\frac{a}{b} = \frac{b'}{c'}.$$

In like manner,
$$\frac{b}{c} = \frac{a'}{b'}.$$

Hence, multiplying
$$\frac{a}{c} = \frac{a'}{c'}.$$

Therefore, if a be greater than c, a' is greater than c'; if equal, equal; if less, less.

PROP. XXII.—Theorem.

If there be two sets of magnitudes, which, taken two by two in direct order, have equal ratios, then the first : the last of the first set :: the first : the last of the second set ("ex aequali," or "ex aequo").

Let a, b, c; a', b', c' be the two sets of magnitudes, and if $a : b :: a' : b'$, and $b : c :: b' : c'$, then $a : c :: a' : c'$.

Dem.—Since
$$a : b :: a' : b',$$

we have
$$\frac{a}{b} = \frac{a'}{b'}$$

In like manner,
$$\frac{b}{c} = \frac{b'}{c'}.$$

Hence, multiplying,
$$\frac{a}{c} = \frac{a'}{c'}.$$

Therefore
$$a : c :: a' : c',$$

and similarly for any number of magnitudes in each set.

Cor. 1.—If the ratio $b : c$ be equal to the ratio $a : b$, then a, b, c will be in continued proportion, and so will a', b', c'. Hence [Def. XII. Annotation 3],

$$\frac{a}{c} = \frac{a^2}{b^2} \text{ and } \frac{a'}{c'} = \frac{a'^2}{b'^2};$$

but
$$\frac{a}{c} = \frac{a'}{c'}. \qquad\qquad\qquad\qquad \text{[XXII.]}$$

Therefore
$$\frac{a^2}{b^2} = \frac{a'^2}{b'^2}.$$

Hence, if
$$a : b :: a' : b',$$
$$a^2 : b^2 :: a'^2 : b'^2$$

Or if four magnitudes be proportional, their squares are proportional.

Cor. 2.—If four magnitudes be proportional, their cubes are proportional.

PROP. XXIII.—THEOREM.

If there be two sets of magnitudes, which, taken two by two in transverse order, have equal ratios; then the first : the last of the first set :: the first : the last of the second set ("ex aequo perturbato").

Let a, b, c; a', b', c' be the two sets of magnitudes, and let the ratio $a : b = b' : c'$, and $b : c = a' : b'$; then $a : c :: a' : c'$.

Dem.—Since
$$a : b :: b' : c',$$

we have
$$\frac{a}{b} = \frac{b'}{c'}.$$

In like manner,
$$\frac{b}{c} = \frac{a'}{b'}.$$

Hence, multiplying,
$$\frac{a}{c} = \frac{a'}{c'};$$

therefore
$$a : c :: a' : c',$$

133

and similarly for any number of magnitudes in each set.

This Proposition and the preceding one may be included in one enunciation, thus: "*Ratios compounded of equal ratios are equal.*"

PROP. XXIV.—THEOREM.

If two magnitudes of the same kind (a, b) have to a third magnitude (c) ratios equal to those which two other magnitudes (a′, b′) have to a third (c′), then the sum (a + b) of the first two has the same ratio to their third (c) which the sum (a′ + b′) of the other two magnitudes has to their third (c′).

Dem.—Since $a : c :: a' : c'$,

we have $\dfrac{a}{c} = \dfrac{a'}{c'}.$

In like manner, $\dfrac{b}{c} = \dfrac{b'}{c'};$

therefore, adding, $\dfrac{a + b}{c} = \dfrac{a' + b'}{c'}.$

Hence $a + b : c :: a' + b' : c'.$

PROP. XXV.—THEOREM.

If four magnitudes of the same kind be proportionals, the sum of the greatest and least is greater than the sum of the other two.

Let $a : b :: c : d$; then, if a be the greatest, d will be the least [XIV. and A]. It is required to prove that $a + d$ is greater than $b + c$.

Dem.—Since $a : b :: c : d,$

$a : c :: b : d$ [*alternando*];

therefore $a : a - c :: b : b - d$ [**E**].;

but a is greater than b (hyp.),

therefore $a - c$ is greater than $b - d$ [XIV.].

Hence $a + d$ is greater than $b + c$.

Questions for Examination on Book V.

1. What is the subject-matter of this book?
2. When is one magnitude said to be a multiple of another?
3. What is a submultiple or measure?
4. What are equimultiples?
5. What is the ratio of two commensurable magnitudes?
6. What is meant by the ratio of incommensurable magnitudes?
7. Give an Illustration of the ratio of incommensurables.
8. What are the terms of a ratio called?

9. What is a ratio of greater inequality?

10. What is a ratio of lesser inequality?

11. What is the product of two ratios called? *Ans.* The ratio compounded of these ratios.

12. What is *duplicate ratio*?

13. What is Euclid's definition of duplicate ratio?

14. Give another definition.

15. Define triplicate ratio.

16. What is proportion? *Ans.* equality of ratios.

17. Give Euclid's definition of proportion.

18. How many ratios in a proportion?

19. What are the Latin terms in use to denote some of the Propositions of Book V.?

20. When is a line divided harmonically?

21. When a line is divided harmonically, what are corresponding pairs of points called? *Ans.* Harmonic conjugates.

22. What are reciprocal ratios?

23. Give one enunciation that will include Propositions XXII., XXIII. of Book V.

Exercises on Book V.

DEF. I.—A ratio whose antecedent is greater than its consequent is called *a ratio of greater inequality;* and a ratio whose antecedent is less than its consequent, *a ratio of lesser inequality.*

DEF. II.—A right line is said to be cut *harmonically* when it is divided internally and externally in any ratios that are equal in magnitude.

1. A ratio of greater inequality is increased by diminishing its terms by the same quantity, and diminished by increasing its terms by the same quantity.

2. A ratio of lesser inequality is diminished by diminishing its terms by the same quantity, and increased by increasing its terms by the same quantity.

3. If four magnitudes be proportionals, the sum of the first and second is to their difference as the sum of the third and fourth is to their difference (*componendo et dividendo*).

4. If two sets of four magnitudes be proportionals, and if we multiply corresponding terms together, the products are proportionals.

5. If two sets of four magnitudes be proportionals, and if we divide corresponding terms, the quotients are proportionals.

6. If four magnitudes be proportionals, their squares, cubes, &c., are proportionals.

7. It two proportions have three terms of one respectively equal to three corresponding terms of the other, the remaining term of the first is equal to the remaining term of the second.

8. If three magnitudes be continual proportionals, the first is to the third as the square of the difference between the first and second is to the square of the difference between the second and third.

9. If a line AB, cut harmonically in C and D, be bisected in O; prove OC, OB, OD are continual proportionals.

10. In the same case, if O' be the middle point of CD; prove $OO'^2 = OB^2 + O'D^2$.

11. And $AB(AC + AD) = 2AC \cdot AD$, or $\dfrac{1}{AC} + \dfrac{1}{AD} = \dfrac{2}{AB}$.

12. And $CD(AD + BD) = 2AD \cdot BD$, or $\dfrac{1}{BD} + \dfrac{1}{AD} = \dfrac{2}{CD}$.

13. And $AB \cdot CD = 2AD \cdot CB$.

135

BOOK VI.

APPLICATION OF THE THEORY OF PROPORTION

DEFINITIONS.

I. *Similar Rectilineal Figures* are those whose several angles are equal, each to each, and whose sides about the equal angles are proportional.

Similar figures agree in shape; if they agree also in size, they are congruent.

1. When the shape of a figure is given, it is said to be given in *species*. Thus a triangle whose angles are given is given in species. Hence similar figures are of the same species.

2. When the size of a figure is given, it is said to be given in *magnitude*; for instance, a square whose side is of given length.

3. When the place which a figure occupies is known, it is said to be given in *position*.

II. A right line is said to be cut at a point in extreme and mean ratio when the whole line is to the greater segment as the greater segment is to the less.

III. If three quantities of the same kind be in continued proportion, the middle term is called a *mean proportional* between the other two.

Magnitudes in continued proportion are also said to be in *geometrical progression*.

IV. If four quantities of the same kind be in continued proportion, the two middle terms are called two *mean proportionals* between the other two.

V. The *altitude* of any figure is the length of the perpendicular from its highest point to its base.

VI. Two corresponding angles of two figures have the sides about them reciprocally proportional, when a side of the first is to a side of the second as the remaining side of the second is to the remaining side of the first.

This is evidently equivalent to saying that a side of the first is to a side of the second in the reciprocal ratio of the remaining side of the first to the remaining side of the second.

PROP. I.—Theorem.

Triangles (ABC, ACD) and parallelograms (EC, CF) which have the same altitude are to one another as their bases (BC, CD).

Dem.—Produce BD both ways, and cut off any number of parts BG, GH, &c., each equal to CB, and any number DK, KL, each equal to CD. Join AG, AH, AK, AL.

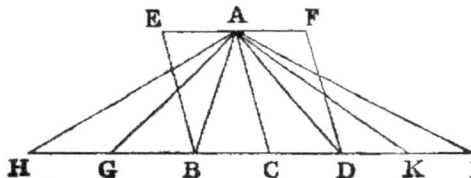

Now, since the several bases CB, BG, GH are all equal, the triangles ACB, ABG, AGH are also all equal [I. XXXVIII.]. Therefore the triangle ACH is the same multiple of ACB that the base CH is of the base CB. In like manner,

136

the triangle ACL is the same multiple of ACD that the base CL is of the base CD; and it is evident that [I. XXXVIII.] if the base HC be greater than CL, the triangle HAC is greater than CAL; if equal, equal; and if less, less. Now we have four magnitudes: the base BC is the first, the base CD the second, the triangle ABC the third, and the triangle ACD the fourth. We have taken equimultiples of the first and third, namely, the base CH, and the triangle ACH; also equimultiples of the second and fourth, namely, the base CL, and the triangle ACL; and we have proved that according as the multiple of the first is greater than, equal to, or less than the multiple of the second, the multiple of the third is greater than, equal to, or less than the multiple of the fourth. Hence [V. Def. V.] *the base $BC : CD ::$ the triangle $ABC : ACD$.*

2. The parallelogram EC is double of the triangle ABC [I. XXXIV.], and the parallelogram CF is double of the triangle ACD. Hence [V. XV.] $EC : CF ::$ the triangle $ABC : ACD$; but $ABC : ACD :: BC : CD$ (Part I.). *Therefore* [V. XI.] $EC : CF ::$ *the base $BC : CD$.*

Or thus: Let A, A' denote the areas of the triangles ABC, ACD, respectively, and P their common altitude; then [II. I., *Cor.* 1],

$$A = \tfrac{1}{2} P . BC, \quad A' = \tfrac{1}{2} P . CD.$$

Hence
$$\frac{A}{A'} = \frac{BC}{CD}, \text{ or } A : A' :: BC : CD.$$

In extending this proof to parallelograms we have only to use P instead of $\tfrac{1}{2} P$.

PROP. II.—THEOREM.

If a line (DE) be parallel to a side (BC) of a triangle (ABC), it divides the remaining sides, measured from the opposite angle (A), proportionally; and, conversely, If two sides of a triangle, measured from an angle, be cut proportionally, the line joining the points of section is parallel to the third side.

1. It is required to prove that $AD : DB :: AE : EC$.

Dem.—Join BE, CD. The triangles BDE, CED are on the same base DE, and between the same parallels BC, DE. Hence [I. XXXVII.] they are equal, and therefore [V. VII.] the triangle $ADE : BDE :: ADE : CDE$;

but $\quad ADE : BDE :: AD : DB$ [I.],

and $\quad ADE : CDE :: AE : EC$ [I.].

Hence $\quad AD : DB :: AE : EC$.

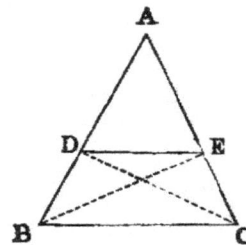

2. If $AD : DB :: AE : EC$, it is required to prove that DE is parallel to BC.

Dem.—Let the same construction be made;

then $\quad AD : DB ::$ the triangle $ADE : BDE$ [I.].

and $\quad AE : EC ::$ the triangle $ADE : CDE$ [I.];

but $\quad AD : DB :: AE : EC$ (hyp.).

Hence $\quad ADE : BDE :: ADE : CDE$.

137

Therefore [V. IX.] the triangle BDE is equal to CDE, and they are on the same base DE, and on the same side of it; hence they are between the same parallels [I. XXXIX.]. *Therefore DE is parallel to BC.*

Observation.—The line DE may cut the sides AB, AC produced through B, C, or through the angle A; but evidently a separate figure for each of these cases is unnecessary.

Exercise.

If two lines be cut by three or more parallels, the intercepts on one are proportional to the corresponding intercepts on the other.

PROP. III.—THEOREM.

If a line (AD) bisect any angle (A) of a triangle (ABC), it divides the opposite side (BC) into segments proportional to the adjacent sides. Conversely, If the segments (BD, DC) into which a line (AD) drawn from any angle (A) of a triangle divides the opposite side be proportional to the adjacent sides, that line bisects the angle (A).

Dem.—1. Through C draw CE parallel to AD, to meet BA produced in E. Because BA meets the parallels AD, EC, the angle BAD [I. XXIX.] is equal to AEC; and because AC meets the parallels AD, EC, the angle DAC is equal to ACE; but the angle BAD is equal to DAC (hyp.); therefore the angle ACE is equal to AEC; therefore AE is equal to AC [I. VI.]. Again, because AD is parallel to EC, one of the sides of the triangle BEC, $BD : DC :: BA : AE$ [II.]; but AE has been proved equal to AC. *Therefore BD : DC :: BA : AC.*

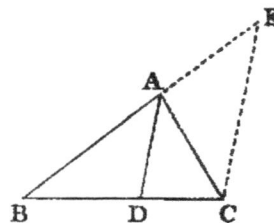

2. If $BD : DC :: BA : AC$, the angle BAC is bisected.

Dem.—Let the same construction be made.

Because AD is parallel to EC, $BA : AE :: BD : DC$ [II.]; but $BD : DC :: BA : AC$ (hyp.). Therefore [V. XI.] $BA : AE :: BA : AC$, and hence [V. IX.] AE is equal to AC; therefore the angle AEC is equal to ACE; but AEC is equal to BAD [I. XXIX.], and ACE to DAC; hence BAD is equal to DAC, *and the line AD bisects the angle BAC.*

Exercises.

1. If the line AD bisect the external vertical angle CAE, $BA : AC :: BD : DC$, and conversely.

Dem.—Cut off $AE = AC$. Join ED. Then the triangles ACD, AED are evidently congruent; therefore the angle EDB is bisected; hence [III.] $BA : AE :: BD : DE$; or $BA : AC :: BD : DC$.

2. Exercise 1 has been proved by quoting Proposition III. Prove it independently, and prove III. as an inference from it.

3. The internal and the external bisectors of the vertical angle of a triangle divide the base harmonically (*see* Definition, p. 191).

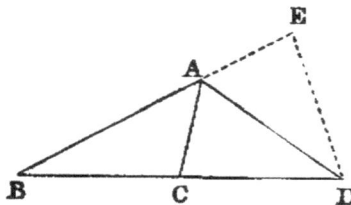

138

4. Any line intersecting the legs of any angle is cut harmonically by the internal and external bisectors of the angle.

5. Any line intersecting the legs of a right angle is cut harmonically by any two lines through its vertex which make equal angles with either of its sides.

6. If the base of a triangle be given in magnitude and position, and the ratio of the sides, the locus of the vertex is a circle which divides the base harmonically in the ratio of the sides.

7. If a, b, c denote the sides of a triangle ABC, and D, D' the points where the internal and external bisectors of A meet BC; prove

$$DD' = \frac{2abc}{b^2 - c^2}.$$

8. In the same case, if E, E', F, F' be points similarly determined on the sides CA, AB, respectively; prove

$$\frac{1}{DD'} + \frac{1}{EE'} + \frac{1}{FF'} = 0,$$

$$and\ \frac{a^2}{DD'} + \frac{b^2}{EE'} + \frac{c^2}{FF'} = 0.$$

PROP. IV.—Theorem.

The sides about the equal angles of equiangular triangles (BAC, CDE) are proportional, and those which are opposite to the equal angles are homologous.

Dem.—Let the sides BC, CE, which are opposite to the equal angles A and D, be conceived to be placed so as to form one continuous line, the triangles being on the same side, and so that the equal angles BCA, CED may not have a common vertex.

Now, the sum of the angles ABC, BCA is less than two right angles; but BCA is equal to BED (hyp.). Therefore the sum of the angles ABE, BED is less than two right angles; hence [I., Axiom XII.] the lines AB, ED will meet if produced. Let them meet in F. Again, because the angle BCA is equal to BEF, the line CA [I. XXVIII.] is parallel to EF. In like manner, BF is parallel to CD; therefore the figure $ACDF$ is a parallelogram; hence AC is equal to DF, and CD is equal to AF. Now, because AC is parallel to FE, $BA : AF :: BC : CE$ [II.]; but AF is equal to CD, therefore $BA : CD :: BC : CE$; hence [V. XVI.]; $AB : BC :: DC : CE$. Again, because CD is parallel to BF, $BC : CE :: FD : DE$; but FD is equal to AC, therefore $BC : CE :: AC : DE$; hence [V. XVI.] $BC : AC :: CE : DE$. Therefore we have proved that $AB : BC :: DC : CE$, and that $BC : CA :: CE : ED$. Hence (*ex aequali*) $AB : AC :: DC : DE$. *Therefore the sides about the equal angles are proportional.*

This Proposition may also be proved very simply by superposition. Thus (*see* fig., Prop. II.): let the two triangles be ABC, ADE; let the second triangle ADE be conceived to be placed on ABC, so that its two sides AD, AE may fall on the sides AB, AC; then, since the angle ADE is equal to ABC, the side DE is parallel to BC. Hence [II.] $AD : DB :: AE : EC$; hence $AD : AB :: AE : AC$, and [V. XVI.] $AD : AE :: AB : AC$. *Therefore the sides about the equal angles BAC, DAE are proportional, and similarly for the others.*

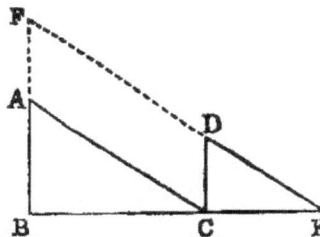

139

It can be proved by this Proposition that two lines which meet at infinity are parallel. For, let I denote the point at infinity through which the two given lines pass, and draw any two parallels intersecting them in the points A, B; A', B'; then the triangles AIB, $A'IB'$ are equiangular; therefore $AI : AB :: A'I : A'B'$; but the first term of the proportion is equal to the third; therefore [V. XIV.] the second term AB is equal to the fourth $A'B'$, and, being parallel to it, the lines AA', BB' [I. XXXIII.] are parallel.

Exercises.

1. If two circles intercept equal chords AB, $A'B'$ on any secant, the tangents AT, $A'T$ to the circles at the points of intersection are to one another as the radii of the circles.

2. If two circles intercept on any secant chords that have a given ratio, the tangents to the circles at the points of intersection have a given ratio, namely, the ratio compounded of the direct ratio of the radii and the inverse ratio of the chords.

3. Being given a circle and a line, prove that a point may be found, such that the rectangle of the perpendiculars let fall on the line from the points of intersection of the circle with any chord through the point shall be given.

4. AB is the diameter of a semicircle ADB; CD a perpendicular to AB; draw through A a chord AF of the semicircle meeting CD in E, so that the ratio $CE : EF$ may be given.

PROP. V.–Theorem.

If two triangles (ABC, DEF) have their sides proportional ($BA : AC :: ED : DF$; $AC : CB :: DF : FE$) they are equiangular, and those angles are equal which are subtended by the homologous sides.

Dem.—At the points D, E make the angles EDG, DEG equal to the angles A, B of the triangle ABC. Then [I. XXXII.] the triangles ABC, DEG are equiangular.

Therefore $\quad BA : AC :: ED : DG$ [IV.];

but $\qquad BA : AC :: ED : DF$ (hyp.).

Therefore DG is equal to DF. In like manner it may be proved that EG is equal to EF. Hence the triangles EDF, EDG have the sides ED, DF in one equal to the sides ED, DG in the other, and the base EF equal to the base EG. Hence [I. VIII.] they are equiangular; but the triangle DEG is equiangular to ABC. *Therefore the triangle DEF is equiangular to ABC.*

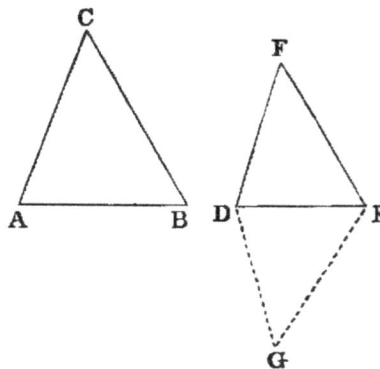

Observation.—In VI. Def. I. two conditions are laid down as necessary for the similitude of rectilineal figures. 1. The equality of angles; 2. The proportionality of sides. Now, from Propositions IV. and V., we see that if two triangles possess either condition, they also possess the other. Triangles are unique in this respect. In all other rectilineal figures one of the conditions may exist without the other. Thus, two quadrilaterals may have their sides proportional without having equal angles, or *vice versâ*.

PROP. VI.—THEOREM.

If two triangles (ABC, DEF) have one angle (A) in one equal to one angle (D) in the other, and the sides about these angles proportional (BA : AC :: ED : DF), the triangles are equiangular, and have those angles equal which are opposite to the homologous sides.

Dem.—Make the same construction as in the last Proposition; then the triangles *ABC*, *DEG* are equiangular.

Therefore $\qquad\qquad BA : AC :: ED : DG$ [IV.];

but $\qquad\qquad\qquad BA : AC :: ED : DF$ (hyp.).

Therefore *DG* is equal to *DF*. Again, because the angle *EDG* is equal to *BAC* (const.), and *BAC* equal to *EDF* (hyp.), the angle *EDG* is equal to *EDF*; and it has been proved that *DG* is equal to *DF*, and *DE* is common; hence the triangles *EDG* and *EDF* are equiangular; but *EDG* is equiangular to *BAC*. *Therefore EDF is equiangular to BAC.*

[It is easy to see, as in the case of Proposition IV., that an immediate proof of this Proposition can also be got from Proposition II.].

Cor. 1.—If the ratio of two sides of a triangle be given, and the angle between them, the triangle is given in species.

PROP. VII.—THEOREM.

If two triangles (ABC, DEF) have one angle (A) one equal to one angle (D) in the other, the sides about two other angles (B, E) proportional (AB : BC :: DE : EF), and the remaining angles (C, F) of the same species (i. e. either both acute or both not acute), the triangles are similar.

Dem.—If the angles *B* and *E* are not equal, one must be greater than the other. Suppose *ABC* to be the greater, and that the part *ABG* is equal to *DEF*, then the triangles *ABG*, *DEF* have two angles in one equal to two angles in the other, and are [I. XXXII.] equiangular.

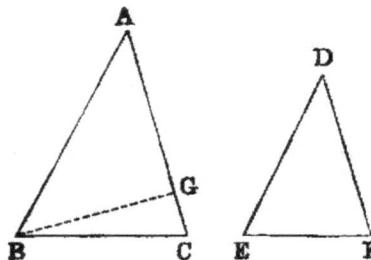

Therefore $\quad AB : BG :: DE : EF$ [IV.];

but $\qquad AB : BC :: DE : EF$ (hyp.).

Therefore *BG* is equal to *BC*. Hence the angles *BCG*, *BGC* must be each acute [I. XVII.]; therefore *AGB* must be obtuse; hence *DFE*, which is equal to it, is obtuse; and it has been proved that *ACB* is acute; therefore the angles *ACB*, *DFE* are of different species; but (hyp.) they are of the same species, which is absurd. Hence the angles *B* and *E* are not unequal, that is, they are equal. *Therefore the triangles are equiangular.*

141

Cor. 1.—If two triangles ABC, DEF have two sides in one proportional to two sides in the other, $AB : BC :: DE : EF$, and the angles A, D opposite one pair of homologous sides equal, the angles C, F opposite the other are either equal or supplemental. This Proposition is nearly identical with VII.

Cor. 2.—If either of the angles C, F be right, the other must be right.

PROP. VIII.—THEOREM.

The triangles (ACD, BCD) *into which a right-angled triangle* (ACB) *is divided, by the perpendicular* (CD) *from the right angle* (C) *on the hypotenuse, are similar to the whole and to one another.*

Dem.—Since the two triangles ADC, ACB have the angle A common, and the angles ADC, ACB equal, each being right, they are [I. XXXII.] equiangular; hence [IV.] they are similar. In like manner it may be proved that BDC is similar to ABC. Hence *ADC, CDB are each similar to ACD, and therefore they are similar to one another.*

Cor. 1.—The perpendicular CD is a mean proportional between the segments AD, DB of the hypotenuse.

For, since the triangles ADC, CDB are equiangular, we have $AD : DC :: DC : DB$; hence DC is a mean proportional between AD, DB (Def. III.).

Cor. 2.—BC is a mean proportional between AB, BD; and AC between AB, AD.

Cor. 3.—The segments AD, DB are in the duplicate of $AC : CB$, or in other words, $AD : DB :: AC^2 : CB^2$,

Cor. 4.—$BA : AD$ in the duplicate ratios of $BA : AC$; and $AB : BD$ in the duplicate ratio of $AB : BC$.

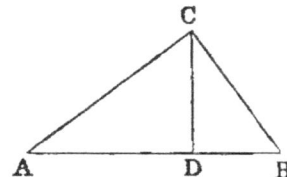

PROP. IX.—PROBLEM.

From a given right line (AB) *to cut off any part required (i.e. to cut off any required submultiple)*

Sol.—Let it be required, for instance, to cut off the fourth part. Draw AF, making any angle with AB, and in AF take any point C, and cut off (I. III.) the parts CD, DE, EF each equal to AC. Join BF, and draw CG parallel to BF. AG is the fourth part of AB.

Dem.—Since CG is parallel to the side BF of the triangle ABF, $AC : AF :: AG : AB$ [II.]; but AC is the fourth part of AF (const.). Hence AG is the fourth part of AB [V., D.]. *In the same manner, any other required submultiple may be cut off.*

Proposition X., Book I., is a particular case of this Proposition.

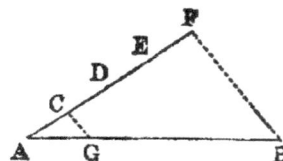

142

PROP. X.—PROBLEM.

To divide a given undivided line (AB) similarly to a given divided line (CD).

Sol.—Draw AG, making any angle with AB, and cut off the parts AH, HI, IG respectively equal to the parts CE, EF, FD of the given divided line CD. Join BG, and draw HK, IL, each parallel to BG. AB will be divided similarly to CD.

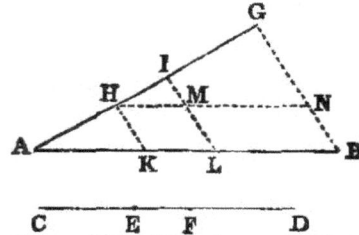

Dem.—Through H draw HN parallel to AB, cutting IL in M. Now in the triangle ALI, HK is parallel to IL. Hence [II.] $AK : KL :: AH : HI$, that is $:: CE : EF$ (const.). Again, in the triangle HNG, MI is parallel to NG. Therefore [II.] $HM : MN :: HI : IG$; but [I. XXXIV.] HM is equal to KL, MN is equal to LB, HI is equal to EF, and IG is equal to FD (const.). Therefore $KL : LB :: EF : FD$. *Hence the line AB is divided similarly to the line CD.*

Exercises.

1. To divide a given line AB *internally* or *externally* in the ratio of two given lines, m, n.

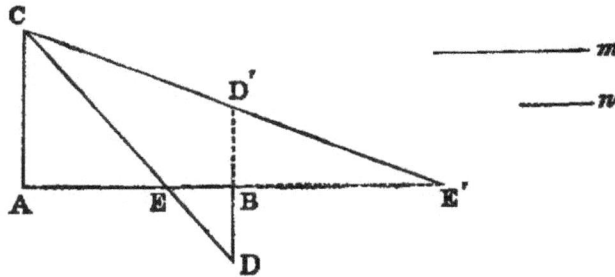

Sol.—Through A and B draw any two parallels AC and BD in opposite directions. Cut off $AC = m$, and $BD = n$, and join CD; the joining line will divide AB internally at E in the ratio of $m : n$.

2. If BD' be drawn in the same direction with AC, as denoted by the dotted line, then CD' will cut AB externally at E' in the ratio of $m : n$.

Cor.—The two points E, E' divide AB harmonically.

This problem is manifestly equivalent to the following:—Given the sum or difference of two lines and their ratio, to find the lines.

3. Any line AE', through the middle point B of the base DD' of a triangle DCD', is cut harmonically by the sides of the triangle and a parallel to the base through the vertex.

4. Given the sum of the squares on two lines and their ratio; find the lines.

5. Given the difference of the squares on two lines and their ratio; find the lines.

6. Given the base and ratio of the sides of a triangle; construct it when any of the following data is given:—1, the area; 2, the difference on the squares of the sides; 3, the sum of the squares on the sides; 4, the vertical angle; 5, the difference of the base angles.

PROP. XI.—Problem.

To find a third proportional to two given lines (X, Y).

Sol.—Draw any two lines AC, AE making an angle. Cut off AB equal X, BC equal Y, and AD equal Y. Join BD, and draw CE parallel to BD, then DE is the third proportional required.

Dem.—In the triangle CAE, BD is parallel to CE; therefore $AB : BC :: AD : DE$ [II.]; but AB is equal to X, and BC, AD each equal to Y. Therefore $X : Y :: Y : DE$. Hence DE is a third proportional to X and Y.

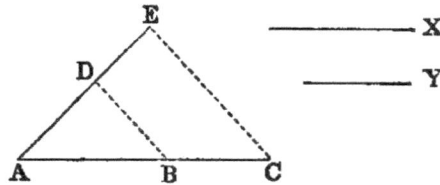

Another solution can be inferred from Proposition VIII. For if AD, DC in that Proposition be respectively equal to X and Y, then DB will be the third proportional. Or again, if in the diagram, Proposition VIII., $AD = X$, and $AC = Y$, AB will be the third proportional. Hence may be inferred a method of continuing the proportion to any number of terms.

Exercises.

1. If $AO\Omega$ be a triangle, having the side $A\Omega$ greater than AO; then if we cut off $AB = AO$, draw BB' parallel to AO, cut off $BC = BB'$, &c., the series of lines AB, BC, CD, &c., are in continual proportion.

2. $AB - BC : AB :: AB : A\Omega$. This is evident by drawing through B' a parallel to $A\Omega$.

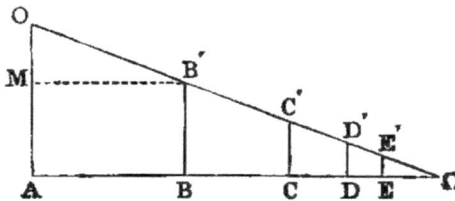

PROP. XII.—Problem.

To find a fourth proportional to three given lines (X, Y, Z).

Sol.—Draw any two lines AC, AE, making an angle; then cut off AB equal X, BC equal Y, AD equal Z. Join BD, and draw CE parallel to BD. *DE will be the fourth proportional required.*

Dem.—Since BD is parallel to CE, we have [II.] $AB : BC :: AD : DE$; therefore $X : Y :: Z : DE$. *Hence DE is a fourth proportional to X, Y, Z.*

Or thus: Take two lines AD, BC intersecting in O. Make $OA = X$, $OB = Y$, $OC = Z$, and describe a circle through the points A, B, C [IV. V.] cutting AD in D. *OD will be the fourth proportional required.*

The demonstration is evident from the similarity of the triangles AOB and COD.

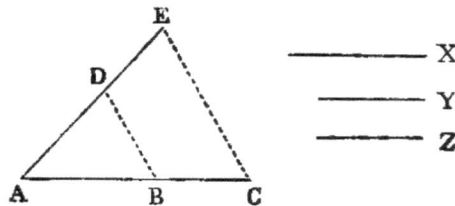

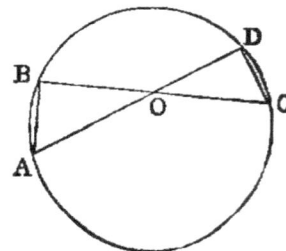

144

PROP. XIII.—PROBLEM.

To find a mean proportional between two given lines. (X, Y).

Sol.—Take on any line AC parts AB, BC respectively equal to X, Y. On AC describe a semicircle ADC. Erect BD at right angles to AC, meeting the semicircle in D. BD will be the mean proportional required.

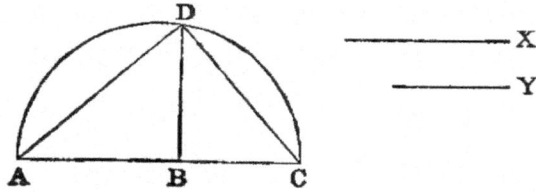

Dem.—Join AD, DC. Since ADC is a semicircle, the angle ADC is right [III. XXXI.]. Hence, since ADC is a right-angled triangle, and BD a perpendicular from the right angle on the hypotenuse, BD is a mean proportional [VIII. *Cor.* 1] between AB, BC; that is, *BD is a mean proportional between X and Y*.

Exercises.

1. Another solution may be inferred from Proposition VIII., *Cor.* 2.

2. If through any point within a circle the chord be drawn, which is bisected in that point, its half is a mean proportional between the segments of any other chord passing through the same point.

3. The tangent to a circle from any external point is a mean proportional between the segments of any secant passing through the same point.

4. If through the middle point C of any arc of a circle any secant be drawn cutting the chord of the arc in D, and the circle again in E, the chord of half the arc is a mean proportional between CD and CE.

5. If a circle be described touching another circle internally and two parallel chords, the perpendicular from the centre of the former on the diameter of the latter, which bisects the chords, is a mean proportional between the two extremes of the three segments into which the diameter is divided by the chords.

6. If a circle be described touching a semicircle and its diameter, the diameter of the circle is a harmonic mean between the segments into which the diameter of the semicircle is divided at the point of contact.

7. State and prove the Proposition corresponding to Ex. 5, for external contact of the circles.

PROP. XIV.—THEOREM.

1. *Equiangular parallelograms* (AB, CD) *which are equal in area have the sides about the equal angles reciprocally proportional—* $AC : CE :: GC : CB$.

2. *Equiangular parallelograms which have the sides about the equal angles reciprocally proportional are equal in area.*

Dem.—Let AC, CE be so placed as to form one right line, and that the equal angles

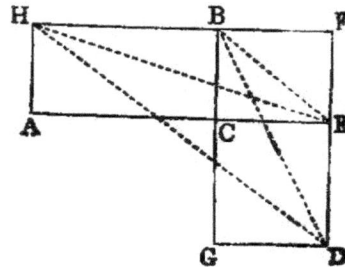

145

ACB, ECG may be vertically opposite. Now, since the angle ACB is equal to ECG, to each add BCE, and we have the sum of the angles ACB, BCE equal to the sum of the angles ECG, BCE; but the sum of ACB, BCE is [I. XIII.] two right angles. Therefore the sum of ECG, BCE is two right angles. Hence [I. XIV.] BC, CG form one right line. Complete the parallelogram BE.

Again, since the parallelograms AB, CD are equal (hyp.),

$$AB : CF :: CD : CF \text{ [V. VII.]};$$
but
$$AB : CF :: AC : CE \text{ [I.]};$$
and
$$CD : CF :: GC : CB \text{ [I.]};$$
therefore
$$AC : CE :: GC : CB;$$

that is, *the sides about the equal angles are reciprocally proportional.*

2. Let $AC : CE :: GC : CB$, to prove the parallelograms AB, CD are equal.

Dem.—Let the same construction be made, we have

$$AB : CF :: AC : CE \text{ [I.]};$$
and
$$CD : CF :: GC : CB \text{ [I.]};$$
but
$$AC : CE :: GC : CB \text{ (hyp.)}.$$
Therefore
$$AB : CF :: CD : CF.$$
Hence
$$AB = CD \quad \text{[V. IX.]};$$

that is, *the parallelograms are equal.*

Or thus: Join HE, BE, HD, BD. The $\square$ HC = twice the $\triangle$ HBE, and the $\square$ CD = twice the $\triangle$ BDE. Therefore the $\triangle$ $HBE = BDE$, and [I. XXXIX.] HD is parallel to BE. Hence

$$HB : BF :: DE : EF; \text{ that is, } AC : CE :: GC : CB.$$

2. May be proved by reversing this demonstration.

Another demonstration of this Proposition may be got by producing the lines HA and DG to meet in I. Then [I. XLIII.] the points I, C, F are collinear, and the Proposition is evident.

PROP. XV.—THEOREM.

1. *Two triangles equal in area* (ACB, DCE), *which have one angle* (C) *in one equal to one angle* (C) *in the other, have the sides about these angles reciprocally proportional.*

2. *Two triangles, which have one angle in one equal to one angle in the other, and the sides about these angles reciprocally proportional, are equal in area.*

Dem.—1. Let the equal angles be so placed as to be vertically opposite, and that AC, CD may form one right line; then it may be demonstrated, as in the last Proposition, that BC, CE form one right line. Join BD.

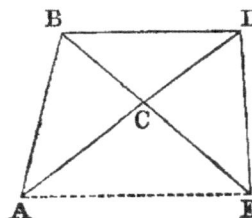

146

Now since the triangles ACB, DCE are equal,

$$ACB : BCD :: DCE : BCD \text{ [V. vii.]};$$
but $\qquad ACB : BCD :: AC : CD \text{ [I.]},$
and $\qquad DCE : BCD :: EC : CB \text{ [I.]}.$
Therefore $\qquad AC : CD :: EC : CB;$

that is, *the sides about the equal angles are reciprocally proportional.*

2. If $AC : CD :: EC : CB$, to prove the triangle ACB equal to DCE.

Dem.—Let the same construction be made, then we have

$$AC : CD :: \text{triangle } ACB : BCD \text{ [I.]},$$
and $\qquad EC : CB :: \text{triangle } DCE : BCD \text{ [I.]};$
but $\qquad AC : CD :: EC : CB \text{ (hyp.)}.$

Therefore the triangle

$$ACB : BCD :: DCE : BCD.$$

Hence the triangle $ACB = DCE$ [V. ix.]—that is, *the triangles are equal.*

This Proposition might have been appended as a *Cor.* to the preceding, since the triangles are the halves of equiangular parallelograms, or it may be proved by joining AE, and showing that it is parallel to BD.

PROP. XVI.—THEOREM.

1. *If four right lines (AB, CD, E, F) be proportional, the rectangle ($AB \cdot F$) contained by the extremes is equal to the rectangle ($CD \cdot E$) contained by the means.*

2. *If the rectangle contained by the extremes of four right lines be equal to the rectangle contained by the means, the four lines are proportional.*

Dem.—1. Erect AH, CI at right angles to AB and CD, and equal to F and E respectively, and complete the rectangles. Then because $AB : CD :: E : F$ (hyp.), and that E is equal to CI, and F to AH (const.), we have $AB : CD :: CI : AH$. Hence the parallelograms AG, CK are equiangular, and have the sides about their equal angles reciprocally proportional. Therefore they are [XIV.] equal; but since AH is equal to F, AG is equal to the rectangle $AB \cdot F$. In like manner, CK is equal to the rectangle $CD \cdot E$. Hence $AB \cdot F = CD \cdot E$; that is, *the rectangle contained by the extremes is equal to the rectangle contained by the means.*

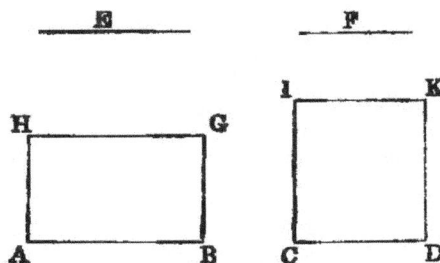

147

2. If $AB . F = CD . E$, to prove $AB : CD :: E : F$.

The same construction being made, because $AB . F = CD . E$, and that F is equal to AH, and E to CI, we have the parallelogram $AG = CK$; and since these parallelograms are equiangular, the sides about their equal angles are reciprocally proportional. Therefore

$$AB : CD :: CI : AH; \text{ that is, } AB : CD :: E : F.$$

Or thus: Place the four lines in a concurrent position so that the extremes may form one continuous line, and the means another. Let the four lines so placed be AO, BO, OD, OC. Join AB, CD. Then because $AO : OB :: OD : OC$, and the angle $AOB = DOC$, the triangles AOB, COD are equiangular. Hence the four points A, B, C, D are concyclic. Therefore [III. XXXV.] $AO . OC = BO . OD$.

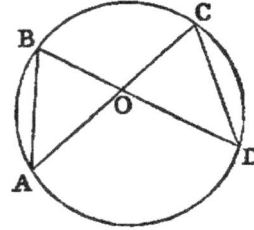

PROP. XVII.—THEOREM

1. *If three right lines (A, B, C) be proportional, the rectangle ($A . C$) contained by the extremes is equal to the square (B^2) of the mean.*

2. *If the rectangle contained by the extremes of three right lines be equal to the square of the mean, the three lines are proportional.*

Dem.—1. Assume a line $D = B$; then because $A : B :: B : C$, we have $A : B :: D : C$. Therefore [XVI.] $AC = BD$; but $BD = B^2$. Therefore $AC = B^2$; that is, *the rectangle contained by the extremes is equal to the square of the mean.*

2. The same construction being made, since $AC = B^2$, we have $A . C = B . D$; therefore $A : B :: D : C$; but $D = B$. Hence $A : B :: B : C$; that is, *the three lines are proportionals.*

This Proposition may be inferred as a *Cor.* to the last, which is one of the fundamental Propositions in Mathematics.

Exercises.

1. If a line CD bisect the vertical angle C of any triangle ACB, its square added to the rectangle $AD . DB$ contained by the segments of the base is equal to the rectangle contained by the sides.

Dem.—Describe a circle about the triangle, and produce CD to meet it in E; then it is easy to see that the triangles ACD, ECB are equiangular. Hence [IV.] $AC : CD :: CE : CB$; therefore $AC . CB = CE . CD = CD^2 + CD . DE = CD^2 + AD . DB$ [III. XXXV.].

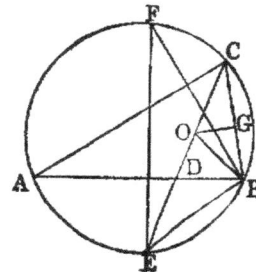

2. If the line CD' bisect the external vertical angle of any triangle ACB, its square subtracted from the rectangle $AD' . D'B$ is equal to $AC . CB$.

3. The rectangle contained by the diameter of the circumscribed circle, and the radius of the inscribed circle of any triangle, is equal to the rectangle contained by the segments of any chord of the circumscribed circle passing through the centre of the inscribed.

148

Dem.—Let O be the centre of the inscribed circle. Join OB (*see* foregoing fig.); let fall the perpendicular OG, draw the diameter EF of the circumscribed circle. Now the angle $ABE = ECB$ [III. XXVII.], and $ABO = OBC$; therefore EBO = sum of OCB, $OBC = EOB$. Hence $EB = EO$. Again, the triangles EBF, OGC are equiangular, because EFB, ECB are equal, and EBF, OGC are each right. Therefore, $EF : EB :: OC : OG$; therefore $EF . OG = EB . OC = EO . OC$.

4. Ex. 3 may be extended to each of the escribed circles of the triangle ACB.

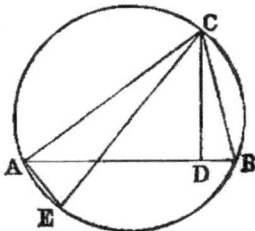

5. The rectangle contained by two sides of a triangle is equal to the rectangle contained by the perpendicular and the diameter of the circumscribed circle. For, let CE be the diameter. Join AE. Then the triangles ACE, DCB are equiangular; hence $AC : CE :: CD : CB$; therefore $AC . CB = CD . CE$.

6. If a circle passing through one of the angles A of a parallelogram $ABCD$ intersect the two sides AB, AD again in the points E, G and the diagonal AC again in F; then $AB . AE + AD . AG = AC . AF$.

Dem.—Join EF, FG, and make the angle $ABH = AFE$. Then the triangles ABH, AFE are equiangular. Therefore $AB : AH :: AF : AE$. Hence $AB . AE = AF . AH$. Again, it is easy to see that the triangles BCH, FAG are equiangular; therefore $BC : CH :: AF : AG$; hence $BC . AG = AF . CH$, or $AD . AG = AF . CH$; but we have proved $AB . AE = AF . AH$. Hence $AD . AG + AB . AE = AF . AC$.

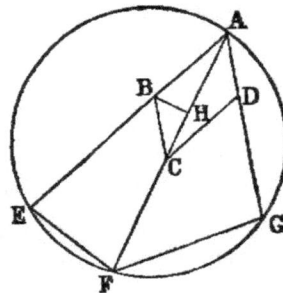

7. If DE, DF be parallels to the sides of a triangle ABC from any point D in the base, then $AB . AE + AC . AF = AD^2 + BD . DC$. This is an easy deduction from 6.

8. If through a point O within a triangle ABC parallels EF, GH, IK to the sides be drawn, the sum of the rectangles of their segments is equal to the rectangle contained by the segments of any chord of the circumscribing circle passing through O.

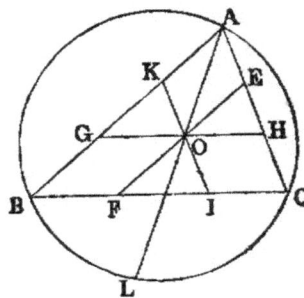

Dem.—$AO . AL = AB . AK + AC . AE.$ (6)

But $AO^2 = AG . AK + AH . AE - GO . OH.$ (7)

Hence $AO . OL = BG . AK + CH . AE + GO . OH,$

or $AO . OL = EO . OF + IO . OK + GO . OH.$

9. The rectangle contained by the side of an inscribed square standing on the base of a triangle, and the sum of the base and altitude, is equal to twice the area of the triangle.

10. The rectangle contained by the side of an escribed square standing on the base of a triangle, and the difference between the base and altitude, is equal to twice the area of the triangle.

149

11. If from any point P in the circumference of a circle a perpendicular be drawn to any chord, its square is equal to the rectangle contained by the perpendiculars from the extremities of the chord on the tangent at P.

12. If O be the point of intersection of the diagonals of a cyclic quadrilateral $ABCD$, the four rectangles $AB.BC$, $BD.CD$, $CD.DA$, $DA.AB$, are proportional to the four lines BO, CO, DO, AO.

13. The sum of the rectangles of the opposite sides of a cyclic quadrilateral $ABCD$ is equal to the rectangle contained by its diagonals.

Dem.—Make the angle $DAO = CAB$; then the triangles DAO, CAB are equiangular; therefore $AD : DO :: AC : CB$; therefore $AD.BC = AC.DO$. Again, the triangles DAC, OAB are equiangular, and $CD : AC :: BO : AB$; therefore $AC.CD = AC.BO$. Hence $AD.BC + AB.CD = AC.BD$.[3]

14. If the quadrilateral $ABCD$ is not cyclic, prove that the three rectangles $AB.CD$, $BC.AD$, $AC.BD$ are proportional to the three sides of a triangle which has an angle equal to the sum of a pair of opposite angles of the quadrilateral.

15. Prove by using Theorem 11 that if perpendiculars be let fall on the sides and diagonals of a cyclic quadrilateral, from any point in the circumference of the circumscribed circle, the rectangle contained by the perpendiculars on the diagonals is equal to the rectangle contained by the perpendiculars on either pair of opposite sides.

16. If AB be the diameter of a semicircle, and PA, PB chords from any point P in the circumference, and if a perpendicular to AB from any point C meet PA, PB in D and E, and the semicircle in F, CF is a mean proportional between CD and CE.

PROP. XVIII.—Problem.

On a given right line (AB) to construct a rectilineal figure similar to a given one $(CDEFG)$, and similarly placed as regards any side (CD) of the latter.

Def.—Similar figures are said to be *similarly described upon given right lines*, when these lines are homologous sides of the figures.

Sol.—Join CE, CF, and construct a triangle ABH on AB equiangular to CDE, and similarly placed as regards CD; that is, make, the angle ABH equal to CDE, and BAH equal to DCE. In like manner construct the triangle HAI

---
[3]This Proposition is known as Ptolemy's theorem.

150

equiangular to ECF, and similarly placed, and lastly, the triangle IAJ equiangular and similarly placed with FCG. *Then $ABHIJ$ is the figure required.*

Dem.—From the construction it is evident that the figures are equiangular, and it is only required to prove that the sides about the equal angles are proportional. Now because the triangle ABH is equiangular to CDE, $AB : BH :: CD : DE$ [IV.]; hence the sides about the equal angles B and D are proportional. Again, from the same triangles we have $BH : HA :: DE : EC$, and from the triangles IHA, FEC; $HA : HI :: EC : EF$; therefore (*ex æquali*) $BH : HI :: DE : EF$; that is, the sides about the equal angles BHI, DEF are proportional, and so in like manner are the sides about the other equal angles. *Hence* (Def. I.) *the figures are similar.*

Observation.—In the foregoing construction, the line AB is homologous to CD, and it is evident that we may take AB to be homologous to any other side of the given figure $CDEFG$. Again, in each case, it the figure $ABHIJ$ be turned round the line AB until it falls on the other side, it will still be similar to the figure $CDEFG$. *Hence on a given line AB there can be constructed two figures each similar to a given figure $CDEFG$, and having the given line AB homologous to any given side CD of the given figure.*

The first of the figures thus constructed is said to be *directly* similar, and the second *inversely* similar to the given figure. These technical terms are due to Hamilton: *see* "Elements of Quaternions," page 112.

Cor. 1.—Twice as many polygons may be constructed on AB similar to a given polygon $CDEFG$ as that figure has sides.

Cor. 2.—If the figure $ABHIJ$ be applied to $CDEFG$ so that the point A will coincide with C, and that the line AB may be placed along CD, then the points H, I, J will be respectively on the lines CE, CF, CG; also the sides BH, HI, IJ of the one polygon will be respectively parallel to their homologous sides DE, EF, FG of the other.

Cor. 3.—If lines drawn from any point O in the plane of a figure to all its angular points be divided in the same ratio, the lines joining the points of division will form a new figure similar to, and having every side parallel to, the homologous side of the original.

PROP. XIX.—Theorem.

Similar triangles (ABC, DEF) have their areas to one another in the duplicate ratio of their homologous sides.

Dem.—Take BG a third proportional to BC, EF [XI.]. Join AG. Then because the triangles ABC, DEF are similar, $AB : BC :: DE : EF$; hence (alternately) $AB : DE :: BC : EF$; but $BC : EF :: EF : BG$ (const.); therefore [V. XI.] $AB : DE :: EF : BG$; hence the sides of the triangles ABG, DEF about the equal angles B, E are reciprocally proportional; therefore the triangles are equal.

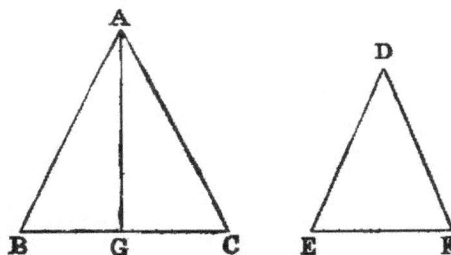

Again, since the lines BC, EF, BG are continual proportionals, $BC : BG$ in the duplicate ratio of $BC : EF$ [V. Def. x.]; but $BC : BG ::$ triangle $ABC : ABG$. Therefore $ABC : ABG$ in the duplicate ratio of $BC : EF$; but it has been proved that the triangle ABG is equal to DEF. *Therefore the triangle ABC is to the triangle DEF in the duplicate ratio of BC : EF.*

This is the first Proposition in Euclid in which the technical term "duplicate ratio" occurs. My experience with pupils is, that they find it very difficult to understand either Euclid's proof or his definition. On this account I submit the following alternative proof, which, however, makes use of a new definition of the duplicate ratio of two lines, viz. the ratio of the squares (*see* Annotations on V. Def. x.) described on these lines.

On AB and DE describe squares, and through C and F draw lines parallel to AB and DE, and complete the rectangles AI, DN.

Now, the triangles JAC, ODF are evidently equiangular.

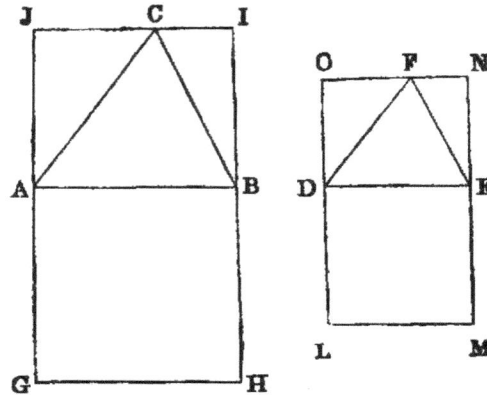

Hence	$JA : AC :: OD : DF$ [iv.];
but	$AC : AB :: DF : DE$ [iv.].
Hence	$JA : AB :: OD : DE$ (*ex æquali*);
but	$AB = AG$, and $DE = DL$;
therefore	$JA : AG :: OD : DL$.
Again,	$JA : AG :: \square AI :$ square AH [i.],
and	$OD : DL :: \square DN :$ square DM [i.].
Hence	$AI : AH :: DN : DM$;
therefore	$AI : DN :: AH : DM$ [V. xvi.];
hence	$\triangle ABC : \triangle DEF :: AB^2 : DE^2$.

152

1. If one of two similar triangles has its sides 50 per cent. longer than the homologous sides of the other; what is the ratio of their areas?

2. When the inscribed and circumscribed regular polygons of any common number of sides to a circle have more than four sides, the difference of their areas is less than the square of the side of the inscribed polygon.

PROP. XX.—THEOREM.

Similar polygons may be divided (1) *into the same number of similar triangles;* (2) *the corresponding triangles have the same ratio to one another which the polygons have;* (3) *the polygons are to each other in the duplicate ratio of their homologous sides.*

Dem.—Let $ABHIJ$, $CDEFG$ be the polygons, and let the sides AB, CD be homologous. Join AH, AI, CE, CF.

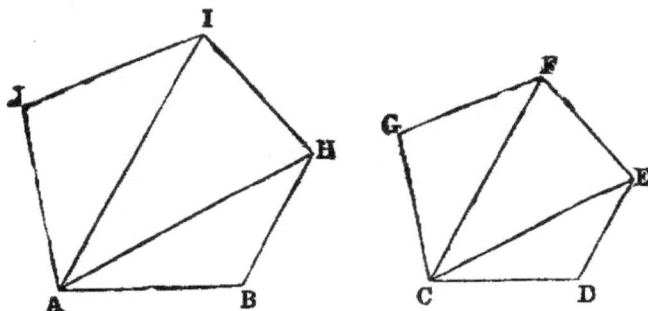

1. The triangles into which the polygons are divided are similar. For, since the polygons are similar, they are equiangular, and have the sides about their equal angles proportional [Def. I.]; hence the angle B is equal to D, and $AB : BH :: CD : DE$; therefore [VI.] the triangle ABH is equiangular to CDE; hence the angle BHA is equal to DEC; but BHI is equal to DEF (hyp.); therefore the angle AHI is equal to CEF. Again, because the polygons are similar, $IH : HB :: FE : ED$; and since the triangles ABH, CDE are similar, $HB : HA :: ED : EC$; hence (*ex aequali*) $IH : HA :: FE : EC$, and the angle IHA has been proved to be equal to the angle FEC; *therefore the triangles IHA, FEC are equiangular.* In the same manner it can be proved that *the remaining triangles are equiangular.*

2. Since the triangle ABH is similar to CDE, we have [XIX.].

$ABH : CDE$ in the duplicate ratio of $AH : CE$.

In like manner,

$AHI : CEF$ in the duplicate ratio of $AH : CE$;

hence $\qquad\qquad ABH : CDE = AHI : CEF$ [V. XI.].

Similarly, $\qquad\qquad AHI : CEF = AIJ : CFG.$

153

In these equal ratios, the triangles ABH, AHI, AIJ are the antecedents, and the triangles CDE, CEF, CFG the consequents, and [V. XII.] any one of these equal ratios is equal to the ratio of the sum of all the antecedents to the sum of all the consequents; *therefore the triangle ABH : the triangle CDE :: the polygon $ABHIJ$: the polygon $CDEFG$.*

3. The triangle ABH : CDE in the duplicate ratio of AB : CD [XIX.]. *Hence (2) the polygon $ABHIJ$: the polygon $CDEFG$ in the duplicate ratio of AB : CD.*

Cor. 1.—The perimeters of similar polygons are to one another in the ratio of their homologous sides.

Cor. 2.—As squares are similar polygons, therefore the duplicate ratio of two lines is equal to the ratio of the squares described on them (compare Annotations, V. Def. X.).

Cor. 3.—Similar portions of similar figures bear the same ratio to each other as the wholes of the figures.

Cor. 4.—Similar portions of the perimeters of similar figures are to each other in the ratio of the whole perimeters.

Exercises.

DEF. I.—*Homologous points* in the planes of two similar figures are such, that lines drawn from them to the angular points of the two figures are proportional to the homologous sides of the two figures.

1. If two figures be similar, to each point in the plane of one there will be a corresponding point in the plane of the other.

Dem.—Let $ABCD$, $A'B'C'D'$ be the two figures, P a point in the plane of $ABCD$. Join AP, BP, and construct a triangle $A'P'B'$ on $A'B'$, similar to APB; then it is easy to see that lines from P' to the angular points of $A'B'C'D'$ are proportional to the lines from P to the angular points of $ABCD$.

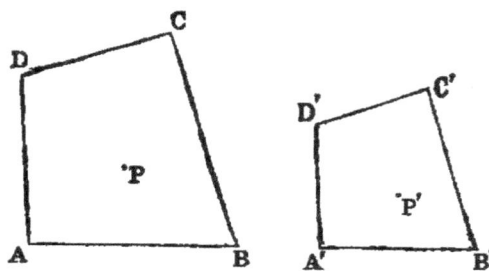

2. If two figures be directly similar, and in the same plane, there is in the plane a special point which, regarded as belonging to either figure, is its own homologous point with respect to the other. For, let AB, $A'B'$ be two homologous sides of the figures, C their point of intersection. Through the two triads of points A, A', C; B, B', C describe two circles intersecting again in the point O: O will be the point required. For it is evident that the triangles OAB, $OA'B'$ are similar and that either may be turned round the point O, so that the two bases, AB, $A'B'$, will be parallel.

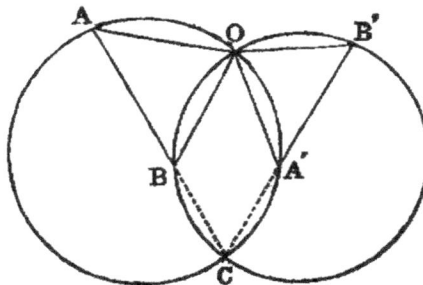

DEF. II.—The point O is called the *centre of similitude of the figures. It is also called their double point.*

3. Two regular polygons of n sides each have n centres of similitude.

154

4. If any number of similar triangles have their corresponding vertices lying on three given lines, they have a common centre of similitude.

5. If two figures be directly similar, and have a pair of homologous sides parallel, every pair of homologous sides will be parallel.

DEF. III.—Two figures, such as those in 5, are said to be *homothetic*.

6. If two figures be homothetic, the lines joining corresponding angular points are concurrent, and the point of concurrence is the centre of similitude of the figures.

7. If two polygons be directly similar, either may be turned round their centre of similitude until they become homothetic, and this may be done in two different ways.

8. Two circles are similar figures.

Dem.—Let O, O' be their centres; let the angle AOB be indefinitely small, so that the arc AB may be regarded as a right line; make the angle $A'O'B'$ equal to AOB; then the triangles AOB, $A'O'B'$ are similar.

Again, make the angle BOC indefinitely small, and make $B'O'C'$ equal to it; the triangles BOC, $B'O'C'$ are similar. Proceeding in this way, we see that the circles can be divided into the same number of similar elementary triangles. Hence the circles are similar figures.

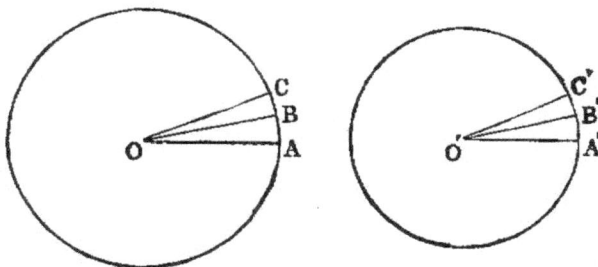

9. Sectors of circles having equal central angles are similar figures.

10. As any two points of two circles may be regarded as homologous, two circles have in consequence an infinite number of centres of similitude; their locus is the circle, whose diameter is the line joining the two points for which the two circles are homothetic.

11. The areas of circles are to one another as the squares of their diameters. For they are to one another as the similar elementary triangles into which they are divided, and these are as the squares of the radii.

12. The circumferences of circles are as their diameters (*Cor.* 1).

13. The circumference of sectors having equal central angles are proportional to their radii. Hence if a, a' denote the arcs of two sectors, which subtend equal angles at the centres, and if r, r' be their radii, $\dfrac{a}{r} = \dfrac{a'}{r'}$.

14. The area of a circle is equal to half the rectangle contained by the circumference and the radius. This is evident by dividing the circle into elementary triangles, as in Ex. 8.

15. The area of a sector of a circle is equal to half the rectangle contained by the arc of the sector and the radius of the circle.

PROP. XXI.—THEOREM.

Rectilineal figures (A, B)*, which are similar to the same figure* (C)*, are similar to one another.*

Dem.—Since the figures A and C are similar, they are equiangular, and have the sides about their equal angles proportional. In like manner B and C are equiangular, and have the sides about their equal angles proportional. Hence A and B are equiangular, and have the sides about their equal angles proportional. *Therefore they are similar.*

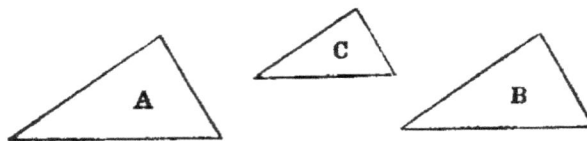

Cor.—Two similar rectilineal figures which are homothetic to a third are homothetic to one another.

Exercise.

If three similar rectilineal figures be homothetic, two by two, their three centres of similitudes are collinear.

PROP. XXII—THEOREM.

If four lines (AB, CD, EF, GH) be proportional, and any pair of similar rectilineal figures (ABK, CDL) be similarly described on the first and second, and also any pair (EI, GJ) on the third and fourth, these figures are proportional. Conversely, if any rectilineal figure described on the first of four right lines: the similar and similarly described figure described on the second :: any rectilineal figure on the third : the similar and similarly described figure on the fourth, the four lines are proportional.

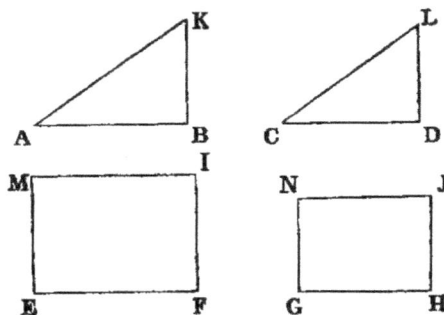

Dem. 1.—$ABK : CDL :: AB^2 : CD^2$. [xx.];

and $\qquad EI : GJ :: EF^2 : GH^2$ [xx.].

But since $\qquad AB : CD :: EF : GH,$

$\qquad AB^2 : CD^2 :: EF^2 : GH^2$ [V. xxii., *Cor.* 1];

therefore $\qquad ABK : CDL :: EI : GJ.$

If $ABK : CDL :: EI : GJ, AB : CD :: EF : GH.$

Dem. 2.—$ABK : CDL :: AB^2 : CD^2 \qquad$ [xx.],

and $\qquad EI : GJ :: EF^2 : GH^2 \qquad$ [xx.];

therefore $\qquad AB^2 : CD^2 :: EF^2 : GH^2.$

Hence $\qquad AB : CD :: EF : GH.$

156

The enunciation of this Proposition is wrongly stated in Simson's *Euclid*, and in those that copy it. As given in those works, the four figures should be similar.

PROP. XXIII.—Theorem.

Equiangular parallelograms (AD, CG) are to each other as the rectangles contained by their sides about a pair of equal angles.

Dem.—Let the two sides AB, BC about the equal angles ABD, CBG, be placed so as to form one right line; then it is evident, as in Prop. XIV., that GB, BD form one right line. Complete the parallelogram BF. Now, denoting the parallelograms AB, BF, CG by X, Y, Z, respectively, we have—

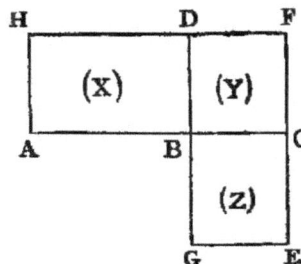

$$X : Y :: AB : BC \quad [\text{I.}],$$
$$Y : Z :: BD : BG \quad [\text{I.}].$$

Hence $\quad XY : YZ :: AB \cdot BD : BC \cdot BG;$

or $\qquad X : Z :: AB \cdot BD : BC \cdot BG.$

Observation.—Since $AB \cdot BD : BC \cdot BG$ is compounded of the two ratios $AB : BC$ and $BD : BG$ [V. Def. of compound ratio], the enunciation is the same as if we said, "in the ratio compounded of the ratios of the sides," which is Euclid's; but it is more easily understood as we have put it.

Exercises.

1. Triangles which have one angle of one equal or supplemental to one angle of the other, are to one another in the ratio of the rectangles of the sides about those angles.

2. Two quadrilaterals whose diagonals intersect at equal angles are to one another in the ratio of the rectangles of the diagonals.

PROP. XXIV.—Theorem.

In any parallelogram (AC), every two parallelograms (AF, FC) which are about a diagonal are similar to the whole and to one another.

Dem.—Since the parallelograms AC, AF have a common angle, they are equiangular [I. XXXIV.], and all that is required to be proved is, that the sides about the equal angles are proportional. Now, since the lines EF, BC are parallel, the triangles AEF, ABC are equiangular; therefore [IV.] $AE : EF :: AB : BC$, and the other sides of the parallelograms are equal to AE, EF; AB, BC: hence the sides about the equal angles are proportional; *therefore the parallelograms AF, AC are similar.* In the same manner *the parallelograms AF, FC are similar.*

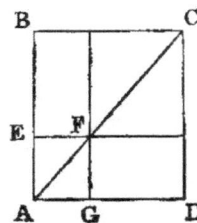

Cor.—The parallelograms AF, FC, AC are, two by two, homothetic.

157

PROP. XXV.—Problem.

To describe a rectilineal figure equal to a given one (A), and similar to another given one (BCD).

Sol.—On any side BC of the figure BCD describe the rectangle BE equal to BCD [I. xlv.], and on CE describe the rectangle EF equal to A. Between BC, CF find a mean proportional GH, and on it describe the figure GHI similar to BCD [xviii.], so that BC and GH may be homologous sides. *GHI is the figure required.*

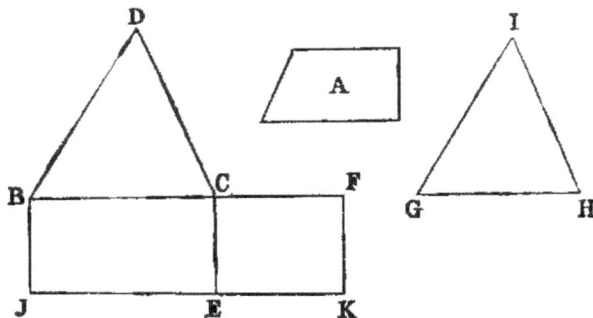

Dem.—The three lines BC, GH, CF are in continued proportion; therefore $BC : CF$ in the duplicate ratio of $BC : GH$ [V. Def. x.]; and since the figures BCD, GHI are similar, $BCD : GHI$ in the duplicate ratio of $BC : GH$ [xx.]; also $BC : CF ::$ rectangle BE : rectangle EF. Hence rectangle $BE : EF ::$ figure $BCD : GHI$; but the rectangle BE is equal to the figure BCD; therefore the rectangle EF is equal to the figure GHI; but EF is equal to A (const.). *Therefore the figure GHI is equal to A, and it is similar to BCD. Hence it is the figure required.*

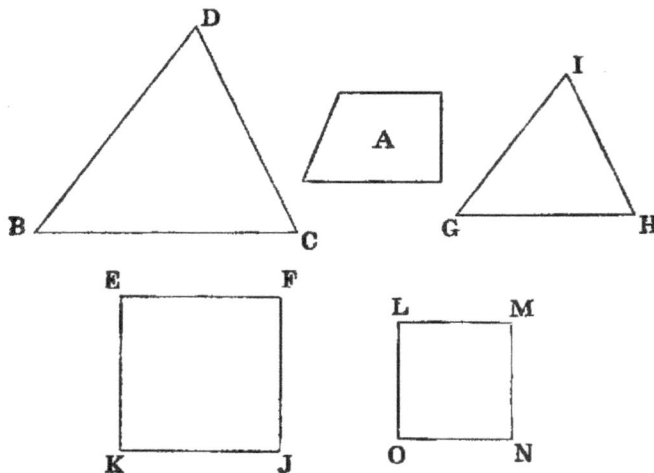

Or thus: Describe the squares *EFJK*, *LMNO* equal to the figures *BCD* and *A* respectively [II. XIV.]; then find *GH* a fourth proportional to *EF*, *LM*, and *BC* [XII.]. On *GH* describe the rectilineal figure *GHI* similar to the figure *BCD* [XVIII.], so that *BC* and *GH* may be homologous sides. *GHI is the figure required.*

Dem.—Because *EF* : *LM* :: *BC* : *GH* (const.), the figure *EFJK* : *LMNO* :: *BCD* : *GHI* [XXII.]; but *EFJK* is equal to *BCD* (const.); therefore *LMNO* is equal to *GHI*; but *LMNO* is equal to *A* (const.). *Therefore GHI is equal to A, and it is similar to BCD.*

PROP. XXVI.—THEOREM.

If two similar and similarly situated parallelograms (AEFG, ABCD) have a common angle, they are about the same diagonal.

Dem.—Draw the diagonals (*see* fig., Prop. XXIV.) *AF*, *AC*. Then because the parallelograms *AEFG*, *ABCD* are similar figures, they can be divided into the same number of similar triangles [XX.]. Hence the triangle *FAG* is similar to *CAD*, and therefore the angle *FAG* is equal to the angle *CAD*. Hence the line *AC* must pass through the point *F*, and *therefore the parallelograms are about the same diagonal.*

Observation.—Proposition XXVI., being the converse of XXIV., has evidently been misplaced. The following would be a simpler enunciation:—"If two homothetic parallelograms have a common angle, they are about the same diagonal."

PROP. XXVII—PROBLEM.

To inscribe in a given triangle (ABC) the maximum parallelogram having a common angle (B) with the triangle.

Sol.—Bisect the side *AC* opposite to the angle *B*, at *P* : through *P* draw *PE*, *PF* parallel to the other sides of the triangle. *BP is the parallelogram required.*

Dem.—Take any other point *D* in *AC* : draw *DG*, *DH* parallel to the sides, and *CK* parallel to *AB*; produce *EP*, *GD* to meet *CK* in *K* and *J*, and produce *HD* to meet *PK* in *I*.

Now, since *AC* is bisected in *P*, *EK* is also bisected in *P*; hence [I. XXXVI.] the parallelogram *EO* is equal to *OK*; therefore *EO* is greater than *DK*; but *DK* is equal to *FD* [I. XLIII.]; hence *EO* is greater than *FD*. To each add *BO*, and we have the parallelogram *BP* greater than *BD*. *Hence BP is the maximum parallelogram which can be inscribed in the given triangle.*

Cor. 1.—The maximum parallelogram exceeds any other parallelogram about the same angle in the triangle, by the area of the similar parallelogram whose diagonal is the line between the middle point *P* of the opposite side and the point *D*, which is the corner of the other inscribed parallelogram.

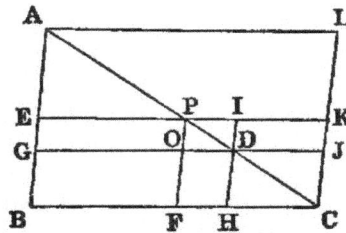

159

Cor. 2.—The parallelograms inscribed in a triangle, and having one angle common with it, are proportional to the rectangles contained by the segments of the sides of the triangle, made by the opposite corners of the parallelograms.

Cor. 3.—The parallelogram $AC : GH :: AC^2 : AD \cdot DC$.

PROP. XXVIII.—PROBLEM.

To inscribe in a given triangle (ABC) a parallelogram equal to a given rectilineal figure (X) not greater than the maximum inscribed parallelogram, and having an angle (B) common with the triangle.

Sol.—Bisect the side AC opposite to B, at P. Draw PF, PE parallel to the sides AB, BC; then [XXVII.] BP is the maximum parallelogram that can be inscribed in the triangle ABC; and if X be equal to it, the problem is solved. If not, produce EP, and draw CJ parallel to PF; then describe the parallelogram $KLMN$ [XXV.] equal to the difference between the figure $PJCF$ and X, and similar to $PJCF$, and so that the sides PJ and KL will be homologous; then cut off PI equal to KL; draw IH parallel to AB, cutting AC in D, and draw DG parallel to BC. *BD is the parallelogram required.*

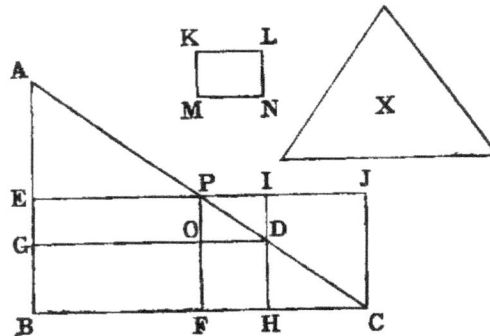

Dem.—Since the parallelograms PC, PD are about the same diagonal, they are similar [XXIV.]; but PC is similar to KPT (const.); therefore PD is similar to KN, and (const.) their homologous sides, PI and KL, are equal; hence [XX.] PD is equal to KN. Now, PD is the difference between EF and GH [XXVII. *Cor.* 1], and KN is (const.) the difference between PC and X; therefore the difference between PC and X is equal to the difference between EF and GH; but EF is equal to PC. *Hence GH is equal to X.*

PROP. XXIX.—PROBLEM.

To escribe to a given triangle (ABC) a parallelogram equal to a given rectilineal figure (X), and having an angle common with an external angle (B) of the triangle.

160

Sol.—The construction is the same as the last, except that, instead of making the parallelogram KN equal to the excess of the parallelogram PC over the rectilineal figure X, we make it equal to their sum; and then make PI equal to KL; draw IH parallel to AB, and the rest of the construction as before.

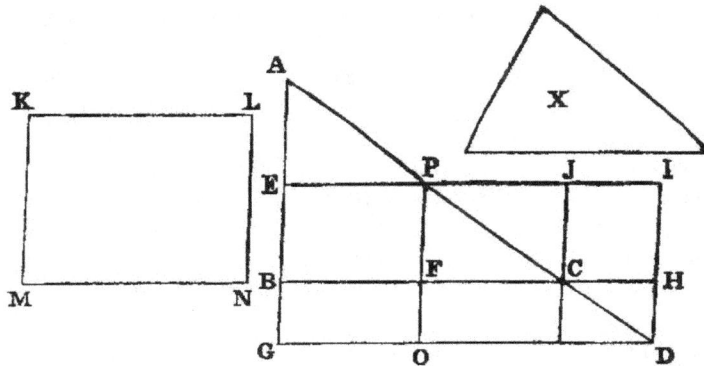

Dem.—Now it can be proved, as in II. VI., that the parallelogram BD is equal to the gnomon OHJ; that is, equal to the difference between the parallelograms PD and PC, or the difference (const.) between KN and PC; that is (const.), equal to X, and BD is escribed to the triangle ABC, and has an angle common with the external angle B. *Hence the thing required is done.*

Observation.—The enunciations of the three foregoing Propositions have been altered, in order to express them in modern technical language. Some writers recommend the student to omit them—we think differently. In the form we have given them they are freed from their usual repulsive appearance. The constructions and demonstrations are Euclid's, but slightly modified.

PROP. XXX.—THEOREM.

To divide a given line (AB) in "extreme and mean ratio."

Sol.—Divide AB in C, so that the rectangle $AB \cdot BC$ may be equal to the square on AC [II. XI.] *Then C is the point required.*

Dem.—Because the rectangle $AB \cdot BC$ is equal to the square on AC, $AB : AC :: AC : BC$ [XVII.]. *Hence AB is cut in extreme and mean ratio in C* [Def. II.].

Exercises.

1. If the three sides of a right-angled triangle be in continued proportion, the hypotenuse is divided in extreme and mean ratio by the perpendicular from the right angle on the hypotenuse.

2. In the same case the greater segment of the hypotenuse is equal to the least side of the triangle.

3. The square on the diameter of the circle described about the triangle formed by the points F, H, D (*see* fig. II. XI.), is equal to six times the square on the line FD.

PROP. XXXI.—THEOREM.

If any similar rectilineal figure be similarly described on the three sides of a right-angled triangle (ABC), the figure on the hypotenuse is equal to the sum of those described on the two other sides.

Dem.—Draw the perpendicular CD [I. XII.]. Then because ABC is a right-angled triangle, and CD is drawn from the right angle perpendicular to the hypotenuse; $BD : AD$ in the duplicate ratio of $BA : AC$ [VIII. *Cor.* 4]. Again, because the figures described on BA, AC are similar, they are in the duplicate ratio of $BA : AC$ [XX.]. Hence [V. XI.] $BA : AD ::$ figure described on BA : figure described on AC. In like manner, $AB : BD ::$ figure described on AB : figure described on BC. Hence [V. XXIV.] AB : sum of AD and $BD ::$ figure described on the line AB : sum of the figures described on the lines AC, BC; but AB is equal to the sum of AD and BD. Therefore [V. A.] *the figure described on the line AB is equal to the sum of the similar figures described on the lines AC and BC.*

Or thus: Let us denote the sides by a, b, c, and the figures by α, β, γ; then because the figures are similar, we have [XX.]

$$\alpha : \gamma :: a^2 : c^2$$

therefore
$$\frac{\alpha}{\gamma} = \frac{a^2}{c^2}.$$

In like manner,
$$\frac{\beta}{\gamma} = \frac{b^2}{c^2};$$

therefore
$$\frac{\alpha + \beta}{\gamma} = \frac{a^2 + b^2}{c^2};$$

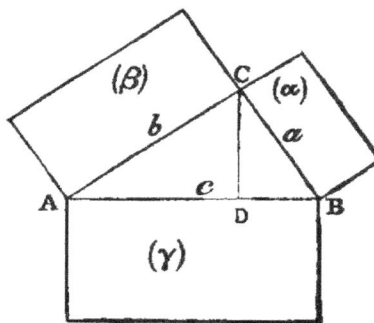

but $a^2 + b^2 = c^2$ [I. XLVII.]. Therefore $\alpha + \beta = \gamma$; that is, *the sum of the figures on the sides is equal to the figure on the hypotenuse.*

Exercise.

If semicircles be described on supplemental chords of a semicircle, the sum of the areas of the two crescents thus formed is equal to the area of the triangle whose sides are the supplemental chords and the diameter.

PROP. XXXII.—THEOREM.

If two triangles (ABC, CDE) which have two sides of one proportional to two sides of the other (AB : BC :: CD : DE), and the contained angles (B, D) equal, be joined at an angle (C), so as to have their homologous sides parallel, the remaining sides are in the same right line.

Dem.—Because the triangles ABC, CDE have the angles B and D equal, and the sides about these angles proportional, viz., $AB : BC :: CD : DE$, they are equiangular [VI.]; therefore the angle BAC is equal to DCE. To each add ACD, and we have the sum of the angles BAC, ACD equal to the sum of DCE and ACD; but the sum of BAC, ACD is [I. XXIX.] two right angles; therefore the sum of DCE and ACD is two right angles. *Hence* [I. XIV.] *AC, CE are in the same right line.*

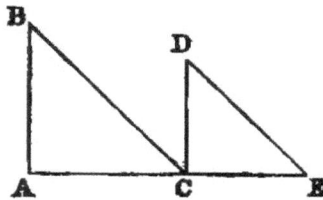

PROP. XXXIII.–THEOREM.

In equal circles, angles $(BOC,\ EPF)$ *at the centres or* $(BAC,\ EDF)$ *at the circumferences have the same ratio to one another as the arcs* $(BC,\ EF)$ *on which they stand, and so also have the sectors* $(BOC,\ EPF)$.

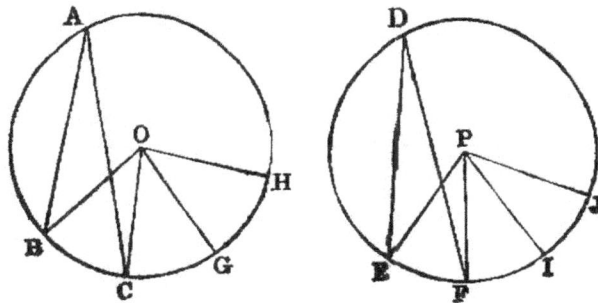

Dem.—1. Take any number of arcs CG, GH in the first circle, each equal to BC. Join OG, OH, and in the second circle take any number of arcs FI, IJ, each equal to EF. Join IP, JP. Then because the arcs BC, CG, GH are all equal, the angles BOC, COG, GOH, are all equal [III. XXVII.]. Therefore the arc BH and the angle BOH are equimultiples of the arc BC and the angle BOC. In like manner it may be proved that the arc EJ and the angle EPJ are equimultiples of the arc EF and the angle EPF. Again, since the circles are equal, it is evident that the angle BOH is greater than, equal to, or less than the angle EPJ, according as the arc BH is greater than, equal to, or less than the arc EJ. Now we have four magnitudes, namely, the arc BC, the arc EF, the angle BOC, and the angle EPF; and we have taken equimultiples of the first and third, namely, the arc BH, the angle BOH, and other equimultiples of the second and fourth, namely, the arc EJ and the angle EPJ, and we have proved that, according as the multiple of the first is greater than, equal to, or less than the multiple of the second, the multiple of the third is greater than, equal to, or less than the multiple of the fourth. *Hence* [V. Def. V.] $BC : EF :: $ *the angle $BOC : EPF$.*

163

Again, since the angle BAC is half the angle BOC [III. xx.], and EDF is half the angle EPF,

$$BAC : EDF :: BOC : EPF \qquad \text{[V. xv.]};$$
but $\qquad BOC : EPF :: BC : EF.$
Hence $\qquad BAC : EDF :: BC : EF \qquad \text{[V. xi.]}.$

2. The sector BOC : sector $EPF :: BC : EF$.

Dem.—The same construction being made, since the arc BC is equal to CG, the angle BOC is equal to COG. Hence the sectors BOC, COG are congruent (*see* Observation, Proposition xxix., Book III.); therefore they are equal. In like manner the sectors COG, GOH are equal. Hence there are as many equal sectors as there are equal arcs; therefore the arc BH and the sector BOH are equimultiples of the arc BC and the sector BOC. In the same manner it may be proved that the arc EJ and the sector EPJ are equimultiples of the arc EF and the sector EPF; and it is evident, by superposition, that if the arc BH is greater than, equal to, or less than the arc EJ, the sector BOH is greater than, equal to, or less than the sector EPJ. *Hence* [V. Def. v.] *the arc $BC : EF ::$ sector BOC : sector EPF.*

The second part may be proved as follows:—

Sector $BOC = \frac{1}{2}$ rectangle contained by the arc BC, and the radius of the circle ABC [xx. Ex. 14] and sector $EPF = \frac{1}{2}$ rectangle contained by the arc EF and the radius of the circle EDF; and since the circles are equal, their radii are equal. Hence, sector BOC : sector $EPF ::$ arc BC : arc EF.

Questions for Examination on Book VI.

1. What is the subject-matter of Book VI.? *Ans.* Application of the theory of proportion.
2. What are similar rectilineal figures?
3. What do similar figures agree in?
4. How many conditions are necessary to define similar triangles?
5. How many to define similar rectilineal figures of more than three sides?
6. When is a figure said to be given in species?
7. When in magnitude?
8. When in position?
9. What is a mean proportional between two lines?
10. Define two mean proportionals.
11. What is the altitude of a rectilineal figure?
12. If two triangles have equal altitudes, how do their areas vary?
13. How do these areas vary if they have equal bases but unequal altitudes?
14. If both bases and altitudes differ, how do the areas vary?
15. When are two lines divided proportionally?
16. If in two lines divided proportionally a pair of homologous points coincide with their point of intersection, what property holds for the lines joining the other pairs of homologous points?
17. Define reciprocal proportion.
18. If two triangles have equal areas, prove that their perpendiculars are reciprocally proportional to the bases.
19. What is meant by figures inversely similar?
20. If two figures be inversely similar, how can they be changed into figures directly similar?

21. Give an example of two triangles inversely similar. *Ans.* If two lines passing through any point O outside a circle intersect it in pairs of points A, A'; B, B', respectively, the triangles OAB, $OA'B'$, are inversely similar.

22. What point is it round which a figure can be turned so as to bring its sides into positions of parallelism with the sides of a similar rectilineal figure. *Ans.* The centre of similitude of the two figures.

23. How many figures similar to a given rectilineal figure of sides can be described on a given line?

24. How many centres of similitude can two regular polygons of n sides each have? *Ans.* n centres, which lie on a circle.

25. What are homothetic figures?

26. How do the areas of similar rectilineal figures vary?

27. What proposition is XIX. a special case of?

28. Define Philo's line.

29. How many centres of similitude have two circles?

Exercises on Book VI.

1. If in a fixed triangle we draw a variable parallel to the base, the locus of the points of intersection of the diagonals of the trapezium thus cut off from the triangle is the median that bisects the base.

2. Find the locus of the point which divides in a given ratio the several lines drawn from a given point to the circumference of a given circle.

3. Two lines AB, XY, are given in position: AB is divided in C in the ratio $m : n$, and parallels AA', BB', CC', are drawn in any direction meeting XY in the points A', B', C'; prove

$$(m + n)CC' = nAA' + mBB'.$$

4. Three concurrent lines from the vertices of a triangle ABC meet the opposite sides in A', B', C'; prove

$$AB' \cdot BC' \cdot CA' = A'B \cdot B'C \cdot C'A.$$

5. If a transversal meet the sides of a triangle ABC in the points A', B', C'; prove

$$AB' \cdot BC' \cdot CA' = -A'B \cdot B'C \cdot C'A.$$

6. If on a variable line AC, drawn from a fixed point A to any point B in the circumference of a given circle, a point C be taken such that the rectangle $AB \cdot AC$ is constant, the locus of C is a circle.

7. If D be the middle point of the base BC of a triangle ABC, E the foot of the perpendicular, L the point where the bisector of the angle A meets BC, H the point of contact of the inscribed circle with BC; prove $DE \cdot HL = HE \cdot HD$.

8. In the same case, if K be the point of contact with BC of the escribed circle, which touches the other sides produced, $LH \cdot BK = BD \cdot LE$.

9. If R, r, r', r'', r''' be the radii of the circumscribed, the inscribed, and the escribed circles of a plane triangle, d, d', d'', d''' the distances of the centre of the circumscribed circle from the centres of the others, then $R^2 = d^2 + 2Rr = d'^2 - 2Rr'$, &c.

10. In the same case, $12R^2 = d^2 + d'^2 + d''^2 + d'''^2$.

11. If p', p'', p''' denote the perpendiculars of a triangle, then

$$(1) \quad \frac{1}{p'} + \frac{1}{p''} + \frac{1}{p'''} = \frac{1}{r};$$

$$(2) \quad \frac{1}{p''} + \frac{1}{p'''} - \frac{1}{p'} = \frac{1}{r'}, \text{ &c.;}$$

$$(3) \quad \frac{2}{p'} = \frac{1}{r} - \frac{1}{r'}, \text{ &c.;}$$

$$(4) \quad \frac{2}{p'} = \frac{1}{r''} + \frac{1}{r'''}, \text{ &c.}$$

12. In a given triangle inscribe another of given form, and having one of its angles at a given point in one of the sides of the original triangle.

13. If a triangle of given form move so that its three sides pass through three fixed points, the locus of any point in its plane is a circle.

14. The angle A and the area of a triangle ABC are given in magnitude: if the point A be fixed in position, and the point B move along a fixed line or circle, the locus of the point C is a circle.

15. One of the vertices of a triangle of given form remains fixed; the locus of another is a right line or circle; find the locus of the third.

16. Find the area of a triangle—(1) in terms of its medians; (2) in terms of its perpendiculars.

17. If two circles touch externally, their common tangent is a mean proportional between their diameters.

18. If there be given three parallel lines, and two fixed points A, B; then if the lines of connexion of A and B to any variable point in one of the parallels intersect the other parallels in the points C and D, E and F, respectively, CF and DE pass each through a fixed point.

19. If a system of circles pass through two fixed points, any two secants passing through one of the points are cut proportionally by the circles.

20. Find a point O in the plane of a triangle ABC, such that the diameters of the three circles, about the triangles OAB, OBC, OCA, may be in the ratios of three given lines.

21. $ABCD$ is a cyclic quadrilateral: the lines AB, AD, and the point C, are given in position; find the locus of the point which divides BD in a given ratio.

22. CA, CB are two tangents to a circle; BE is perpendicular to AD, the diameter through A; prove that CD bisects BE.

23. If three lines from the vertices of a triangle ABC to any interior point O meet the opposite sides in the points A', B', C'; prove

$$\frac{OA'}{AA'} + \frac{OB'}{BB'} + \frac{OC'}{CC'} = 1.$$

24. If three concurrent lines OA, OB, OC be cut by two transversals in the two systems of points A, B, C; A', B', C', respectively: prove

$$\frac{AB}{A'B'} \cdot \frac{OC}{OC'} = \frac{BC}{B'C'} \cdot \frac{OA}{OA'} = \frac{CA}{C'A'} c \frac{OB}{OB'}.$$

25. The line joining the middle points of the diagonals of a quadrilateral circumscribed to a circle—

(1) divides each pair of opposite sides into inversely proportional segments;

(2) is divided by each pair of opposite lines into segments which, measured from the centre, are proportional to the sides;

(3) is divided by both pairs of opposite sides into segments which, measured from either diagonal, have the same ratio to each other.

26. If CD, CD' be the internal and external bisectors of the angle C of the triangle ACB, the three rectangles $AD.DB$, $AC.CB$, $AD.BD'$ are proportional to the squares of AD, AC, AD'; and are—(1) in arithmetical progression if the difference of the base angles be equal to a right angle; (2) in geometrical progression if one base angle be right; (3) in harmonical progression if the sum of the base angles be equal to a right angle.

27. If a variable circle touch two fixed circles, the chord of contact passes through a fixed point on the line connecting the centres of the fixed circles.

166

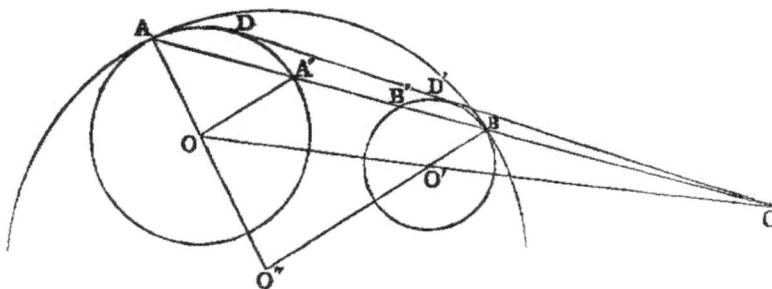

Dem.—Let O, O' be the centres of the two fixed circles; O'' the centre of the variable circle; A, B the points of contact. Let AB and OO' meet in C, and cut the fixed circles again in the points A', B' respectively. Join $A'O$, AO, BO'. Then AO, BO' meet in O'' [III. XI.]. Now, because the triangles OAA', $O''AB$ are isosceles, the angle $O''BA = O''AB = OA'A$. Hence OA' is parallel to $O'B$; therefore $OC : O'C :: OA' : O'B$; that is, in a given ratio. Hence C is a given point.

28. If DD' be the common tangent to the two circles, $DD'^2 = AB' \cdot A'B$.

29. If R denote the radius of O'' and ρ, ρ', the radii of O, O', $DD'^2 : AB^2 :: (R \pm \rho)(R \pm \rho') : R^2$, the choice of sign depending on the nature of the contacts. This follows from 28.

30. If four circles be tangential to a fifth, and if we denote by $\overline{12}$ the common tangent to the first and second, &c., then

$$\overline{12} \cdot \overline{34} + \overline{23} \cdot \overline{14} = \overline{13} \cdot \overline{24}.$$

31. The inscribed and escribed circles of any triangle are all touched by its nine-points circle.

32. The four triangles which are determined by four points, taken three by three, are such that their nine-points circles have one common point.

33. If a, b, c, d denote the four sides, and D, D' the diagonals of a quadrilateral; prove that the sides of the triangle, formed by joining the feet of the perpendiculars from any of its angular points on the sides of the triangle formed by the three remaining points, are proportional to the three rectangles ac, bd, DD'.

34. Prove the converse of Ptolemy's theorem (*see* XVII., Ex. 13).

35. Describe a circle which shall—(1) pass through a given point, and touch two given circles; (2) touch three given circles.

36. If a variable circle touch two fixed circles, the tangent to it from their centre of similitude, through which the chord of contact passes (27), is of constant length.

37. If the lines AD, BD' (*see* fig., Ex. 27) be produced, they meet in a point on the circumference of O'', and the line $O''P$ is perpendicular to DD'.

38. If A, B be two fixed points on two lines given in position, and A', B' two variable points, such that the ratio $AA' : BB'$ is constant, the locus of the point dividing $A'B'$ in a given ratio is a right line.

39. If a line EF divide proportionally two opposite sides of a quadrilateral, and a line GH the other sides, each of these is divided by the other in the same ratio as the sides which determine them.

40. In a given circle inscribe a triangle, such that the triangle whose angular points are the feet of the perpendiculars from the extremities of the base on the bisector of the vertical angle, and the foot of the perpendicular from the vertical angle on the base, may be a maximum.

41. In a circle, the point of intersection of the diagonals of any inscribed quadrilateral coincides with the point of intersection of the diagonals of the circumscribed quadrilateral, whose sides touch the circle at the angular points of the inscribed quadrilateral.

42. Through two given points describe a circle whose common chord with another given circle may be parallel to a given line, or pass through a given point.

43. Being given the centre of a circle, describe it so as to cut the legs of a given angle along a chord parallel to a given line.

167

44. If concurrent lines drawn from the angles of a polygon of an odd number of sides divide the opposite sides each into two segments, the product of one set of alternate segments is equal to the product of the other set.

45. If a triangle be described about a circle, the lines from the points of contact of its sides with the circle to the opposite angular points are concurrent.

46. If a triangle be inscribed in a circle, the tangents to the circle at its three angular points meet the three opposite sides at three collinear points.

47. The external bisectors of the angles of a triangle meet the opposite sides in three collinear points.

48. Describe a circle touching a given line at a given point, and cutting a given circle at a given angle.

DEF.—*The centre of mean position of any number of points A, B, C, D, &c., is a point which may be found as follows*:—Bisect the line joining any two points A, B, in G. Join G to a third point C; divide GC in H, so that $GH = \frac{1}{3}GC$. Join H to a fourth point D, and divide HD in K, so that $HK = \frac{1}{4}HD$, and so on. The last point found will be the centre of mean position of the given points.

49. The centre of mean position of the angular points of a regular polygon is the centre of figure of the polygon.

50. The sum of the perpendiculars let fall from any system of points A, B, C, D, &c., whose number is n on any line L, is equal to n times the perpendicular from the centre of mean position on L.

51. The sum of the squares of lines drawn from any system of points A, B, C, D, &c., to any point P, exceeds the sum of the squares of lines from the same points to their centre of mean position, O, by nOP^2.

52. If a point be taken within a triangle, so as to be the centre of mean position of the feet of the perpendiculars drawn from it to the sides of the triangle, the sum of the squares of the perpendiculars is a minimum.

53. Construct a quadrilateral, being given two opposite angles, the diagonals, and the angle between the diagonals.

54. A circle rolls inside another of double its diameter; find the locus of a fixed point in its circumference.

55. Two points, C, D, in the circumference of a given circle are on the same side of a given diameter; find a point P in the circumference at the other side of the given diameter, AB, such that PC, PD may cut AB at equal distances from the centre.

56. If the sides of any polygon be cut by a transversal, the product of one set of alternate segments is equal to the product of the remaining set.

57. A transversal being drawn cutting the sides of a triangle, the lines from the angles of the triangle to the middle points of the segments of the transversal intercepted by those angles meet the opposite sides in collinear points.

58. If lines be drawn from any point P to the angles of a triangle, the perpendiculars at P to these lines meet the opposite sides of the triangle in three collinear points.

59. Divide a given semicircle into two parts by a perpendicular to the diameter, so that the radii of the circles inscribed in them may have a given ratio.

60. From a point within a triangle perpendiculars are let fall on the sides; find the locus of the point, when the sum of the squares of the lines joining the feet of the perpendiculars is given.

61. If a circle make given intercepts on two fixed lines, the rectangle contained by the perpendiculars from its centre on the bisectors of the angle formed by the lines is given.

62. If the base and the difference of the base angles of a triangle be given, the rectangle contained by the perpendiculars from the vertex on two lines through the middle point of the base, parallel to the internal and external bisectors of the vertical angle, is constant.

63. The rectangle contained by the perpendiculars from the extremities of the base of a triangle, on the internal bisector of the vertical angle, is equal to the rectangle contained by the external bisector and the perpendicular from the middle of the base on the internal bisector.

64. State and prove the corresponding theorem for perpendiculars on the external bisector.

168

65. If R, R' denote the radii of the circles inscribed in the triangles into which a right-angled triangle is divided by the perpendicular from the right angle on the hypotenuse; then, if c be the hypotenuse, and s the semiperimeter, $R^2 + R'^2 = (s - c)^2$.

66. If A, B, C, D be four collinear points, find a point O in the same line with them such that $OA \cdot OD = OB \cdot OC$.

67. The four sides of a cyclic quadrilateral are given; construct it.

68. Being given two circles, find the locus of a point such that tangents from it to the circles may have a given ratio.

69. If four points A, B, C, D be collinear, find the locus of the point P at which AB and CD subtend equal angles.

70. If a circle touch internally two sides, CA, CB, of a triangle and its circumscribed circle, the distance from C to the point of contact on either side is a fourth proportional to the semiperimeter, and CA, CB.

71. State and prove the corresponding theorem for a circle touching the circumscribed circle externally and two sides produced.

72. *Pascal's Theorem.*—If the opposite sides of an irregular hexagon $ABCDEF$ inscribed in a circle be produced till they meet, the three points of intersection G, H, I are collinear.

Dem.—Join AD. Describe a circle about the triangle ADI, cutting the lines AF, CD produced, if necessary, in K and L. Join IK, KL, LI. Now, the angles KLG, FCG are each [III. xxi.] equal to the angle GAD. Hence they are equal. Therefore KL is parallel to CF. Similarly, LI is parallel to CH, and KI to FH; hence the triangles KLI, FCH are homothetic. Hence the lines joining corresponding vertices are concurrent. Therefore the points I, H, G are collinear.

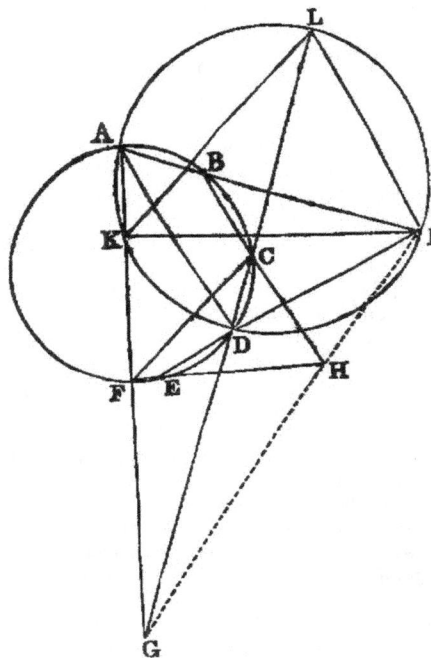

73. If two sides of a triangle circumscribed to a given circle be given in position, but the third side variable, the circle described about the triangle touches a fixed circle.

74. If two sides of a triangle be given in position, and if the area be given in magnitude, two points can be found, at each of which the base subtends a constant angle.

75. If a, b, c, d denote the sides of a cyclic quadrilateral, and s its semiperimeter, prove its area $= \sqrt{(s - a)(s - b)(s - c)(s - d)}$.

76. If three concurrent lines from the angles of a triangle ABC meet the opposite side in the points A', B', C', and the points A', B', C' be joined, forming a second triangle $A'B'C'$,

$$\triangle ABC : \triangle A'B'C' :: AB \cdot BC \cdot CA : 2AB' \cdot BC' \cdot CA'.$$

77. In the same case the diameter of the circle circumscribed about the triangle $ABC = AB' \cdot BC' \cdot CA'$ divided by the area of $A'B'C'$.

78. If a quadrilateral be inscribed in one circle, and circumscribed to another, the square of its area is equal to the product of its four sides.

79. If on the sides AB, AC of a triangle ABC we take two points D, E, and on their line of connexion F, such that

$$\frac{BD}{AD} = \frac{AE}{CE} = \frac{DF}{EF};$$

prove the triangle $BFC = 2ADE$.

169

80. If through the middle points of each of the two diagonals of a quadrilateral we draw a parallel to the other, the lines drawn from their points of intersection to the middle points of the sides divide the quadrilateral into four equal parts.

81. CE, DF are perpendiculars to the diameter of a semicircle, and two circles are described touching CE, DE, and the semicircle, one internally and the other externally; the rectangle contained by the perpendiculars from their centres on AB is equal to $CE \cdot DF$.

82. If lines be drawn from any point in the circumference of a circle to the angular points of any inscribed regular polygon of an odd number of sides, the sums of the alternate lines are equal.

83. If at the extremities of a chord drawn through a given point within a given circle tangents be drawn, the sum of the reciprocals of the perpendiculars from the point upon the tangents is constant.

84. If a cyclic quadrilateral be such that three of its sides pass through three fixed collinear points, the fourth side passes through a fourth fixed point, collinear with the three given ones.

85. If all the sides of a polygon be parallel to given lines, and if the loci of all the angles but one be right lines, the locus of the remaining angle is also a right line.

86. If the vertical angle and the bisector of the vertical angle be given, the sum of the reciprocals of the containing sides is constant.

87. If P, P' denote the areas of two regular polygons of any common number of sides, inscribed and circumscribed to a circle, and Π, Π' the areas of the corresponding polygons of double the number of sides; prove Π is a geometric mean between P and P', and Π' a harmonic mean between Π and P'.

88. The difference of the areas of the triangles formed by joining the centres of the circles described about the equilateral triangles constructed—(1) outwards; (2) inwards—on the sides of any triangle, is equal to the area of that triangle.

89. In the same case, the sum of the squares of the sides of the two new triangles is equal to the sum of the squares of the sides of the original triangle.

90. If R, r denote the radii of the circumscribed and inscribed circles to a regular polygon of any number of sides, R', r', corresponding radii to a regular polygon of the same area, and double the number of sides; prove

$$R' = \sqrt{Rr}, \text{ and } r' = \sqrt{\frac{r(R+r)}{2}}.$$

91. If the altitude of a triangle be equal to its base, the sum of the distances of the orthocentre from the base and from the middle point of the base is equal to half the base.

92. In any triangle, the radius of the circumscribed circle is to the radius of the circle which is the locus of the vertex, when the base and the ratio of the sides are given, as the difference of the squares of the sides is to four times the area.

93. Given the area of a parallelogram, one of its angles, and the difference between its diagonals; construct the parallelogram.

94. If a variable circle touch two equal circles, one internally and the other externally, and perpendiculars be let fall from its centre on the transverse tangents to these circles, the rectangle of the intercepts between the feet of these perpendiculars and the intersection of the tangents is constant.

95. Given the base of a triangle, the vertical angle, and the point in the base whose distance from the vertex is equal half the sum of the sides; construct the triangle.

96. If the middle point of the base BC of an isosceles triangle ABC be the centre of a circle touching the equal sides, prove that any variable tangent to the circle will cut the sides in points D, E, such that the rectangle $BD \cdot CE$ will be constant.

97. Inscribe in a given circle a trapezium, the sum of whose opposite parallel sides is given, and whose area is given.

98. Inscribe in a given circle a polygon all whose sides pass through given points.

99. If two circles X, Y be so related that a triangle may be inscribed in X and circumscribed about Y, an infinite number of such triangles can be constructed.

100. In the same case, the circle inscribed in the triangle formed by joining the points of contact on Y touches a given circle.

101. And the circle described about the triangle formed by drawing tangents to X, at the angular points of the inscribed triangle, touches a given circle.

170

102. Find a point, the sum of whose distances from three given points may be a minimum.

103. A line drawn through the intersection of two tangents to a circle is divided harmonically by the circle and the chord of contact.

104. To construct a quadrilateral similar to a given one whose four sides shall pass through four given points.

105. To construct a quadrilateral, similar to a given one, whose four vertices shall lie on four given lines.

106. Given the base of a triangle, the difference of the base angles, and the rectangle of the sides; construct the triangle.

107. $ABCD$ is a square, the side CD is bisected in E, and the line EF drawn, making the angle $AEF = EAB$; prove that EF divides the side BC in the ratio of $2 : 1$.

108. If any chord be drawn through a fixed point on a diameter of a circle, and its extremities joined to either end of the diameter, the joining lines cut off, on the tangent at the other end, portions whose rectangle is constant.

109. If two circles touch, and through their point of contact two secants be drawn at right angles to each other, cutting the circles respectively in the points A, A'; B, B'; then $AA'^2 + BB'^2$ is constant.

110. If two secants at right angles to each other, passing through one of the points of intersection of two circles, cut the circles again, and the line through their centres in the two systems of points a, b, c; a', b', c' respectively, then $ab : bc :: a'b' : b'c'$.

111. Two circles described to touch an ordinate of a semicircle, the semicircle itself, and the semicircles on the segments of the diameter, are equal to one another.

112. If a chord of a given circle subtend a right angle at a given point, the locus of the intersection of the tangents at its extremities is a circle.

113. The rectangle contained by the segments of the base of a triangle, made by the point of contact of the inscribed circle, is equal to the rectangle contained by the perpendiculars from the extremities of the base on the bisector of the vertical angle.

114. If O be the centre of the inscribed circle of the triangle prove

$$\frac{OA^2}{bc} + \frac{OB^2}{ca} + \frac{OC^2}{ab} = 1.$$

115. State and prove the corresponding theorems for the centres of the escribed circles.

116. Four points A, B, C, D are collinear; find a point P at which the segments AB, BC, CD subtend equal angles.

117. The product of the bisectors of the three angles of a triangle whose sides are a, b, c, is

$$\frac{8abc.s.\text{area}}{(a+b)(b+c)(c+a)}.$$

118. In the same case the product of the alternate segments of the sides made by the bisectors of the angles is

$$\frac{a^2b^2c^2}{(a+b)(b+c)(c+a)}.$$

119. If three of the six points in which a circle meets the sides of any triangle be such, that the lines joining them to the opposite vertices are concurrent, the same property is true of the three remaining points.

120. If a triangle $A'B'C'$ be inscribed in another ABC, prove

$$AB' . BC' . CA' + A'B . B'C . C'A$$

is equal twice the triangle $A'B'C'$ multiplied by the diameter of the circle ABC.

121. Construct a polygon of an odd number of sides, being given that the sides taken in order are divided in given ratios by fixed points.

122. If the external diagonal of a quadrilateral inscribed in a given circle be a chord of another given circle, the locus of its middle point is a circle.

123. If a chord of one circle be a tangent to another, the line connecting the middle point of each arc which it cuts off on the first, to its point of contact with the second, passes through a given point.

124. From a point P in the plane of a given polygon perpendiculars are let fall on its sides; if the area of the polygon formed by joining the feet of the perpendiculars be given, the locus of P is a circle.

BOOK XI.

THEORY OF PLANES, COPLANAR LINES, AND SOLID ANGLES

DEFINITIONS.

I. When two or more lines are in one plane they are said to be *coplanar*.

II. The angle which one plane makes with another is called a *dihedral* angle.

III. A *solid angle* is that which is made by more than two plane angles, in different planes, meeting in a point.

IV. The point is called the *vertex* of the solid angle.

V. If a solid angle be composed of three plane angles it is called a *trihedral angle;* if of four, a *tetrahedral angle;* and if of more than four, a *polyhedral angle.*

PROP. I.—THEOREM.

One part (AB) of a right line cannot be in a plane (X), and another part (BC) not in it.

Dem.—Since *AB* is in the plane *X*, it can be produced in it [Bk. I. Post. II.]; let it be produced to *D*. Then, if *BC* be not in *X*, let any other plane passing through *AD* be turned round *AD* until it passes through the point *C*. Now, because the points *B*, *C* are in this second plane, the line *BC* [I., Def. VI.] is in it. Therefore the two right lines *ABC*, *ABD* lying in one plane have a common segment *AB*, which is impossible. Therefore, &c.

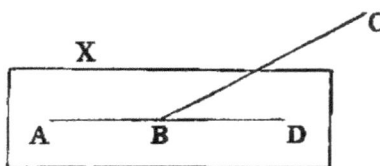

PROP. II.—THEOREM.

Two right lines (AB, CD) which intersect one another in any point (E) are coplanar, and so also are any three right lines (EC, CB, BE) which form a triangle.

Dem.—Let any plane pass through *EB*, and be turned round it until it passes through *C*. Then because the points *E*, *C* are in this plane, the right line *EC* is in it [I., Def. VI.]. For the same reason the line *BC* is in it. *Therefore the lines EC, CB, BE are coplanar;* but *AB* and *CD* are two of these lines. *Hence AB and CD are coplanar.*

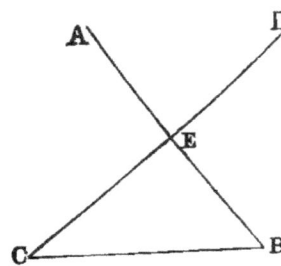

PROP. III.—Theorem.

If two planes (AB, BC) cut one another, their common section (BD) is a right line.

Dem.—If not from B to D, draw in the plane AB the right line BED, and in the plane BC the right line BFD. Then the right lines BED, BFD enclose a space, which [I., Axiom x.] is impossible. *Therefore the common section BD of the two planes must be a right line.*

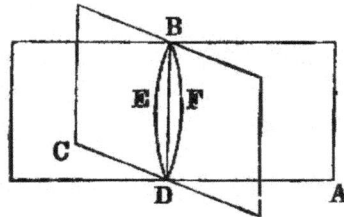

PROP. IV.—Theorem.

If a right line (EF) be perpendicular to each of two intersecting lines (AB, CD), it will be perpendicular to any line GH, which is both coplanar and concurrent with them.

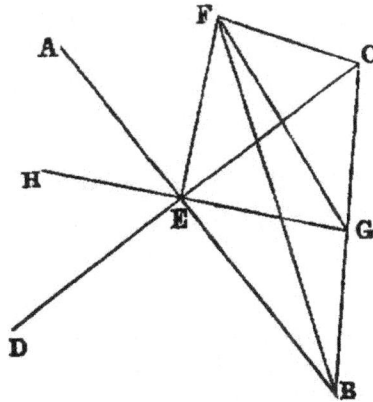

Dem.—Through any point G in GH draw a line BC intersecting AB, CD, and so as to be bisected in G; and join any point F in EF to B, G, C. Then, because EF is perpendicular to the lines EB, EC, we have

$$BF^2 = BE^2 + EF^2, \quad \text{and} \quad CF^2 = CE^2 + EF^2;$$
$$\therefore BF^2 + CF^2 = BE^2 + CE^2 + 2EF^2.$$

Again $\qquad BF^2 + CF^2 = 2BG^2 + 2GF^2$ [II. x. Ex. 2],

and $\qquad BE^2 + CE^2 = 2BG^2 + 2GE^2;$

$$\therefore 2BG^2 + 2GF^2 = 2BG^2 + 2GE^2 + 2EF^2;$$
$$\therefore GF^2 = GE^2 + EF^2.$$

173

Hence the angle GEF is right, and EF is perpendicular to EG.

DEF. VI.—*A line such as EF, which is perpendicular to a system of concurrent and coplanar lines, is said to be perpendicular to the plane of these lines, and is called a* NORMAL *to it.*

Cor. 1.—The normal is the least line that may be drawn from a given point to a given plane; and of all others that may be drawn to it, the lines of any system making equal angles with the normal are equal to each other.

Cor. 2.—A perpendicular to each of two intersecting lines is normal to their plane.

PROP. V.—THEOREM.

If three concurrent lines (BC, BD, BE) have a common perpendicular (AB), they are coplanar.

Dem.—For if possible let BC be not coplanar with BD, BE, and let the plane of AB, BC intersect the plane of BD, BE in the line BF. Then [XI. III.] BF is a right line; and, since it is coplanar with BD, BE, which are each perpendicular to AB, it is [XI. IV.] perpendicular to AB. Therefore the angle ABF is right; and the angle ABC is right (hyp.). Hence ABC is equal to ABF, which is impossible [I., Axiom IX.]. *Therefore the lines BC, BD, BE are coplanar.*

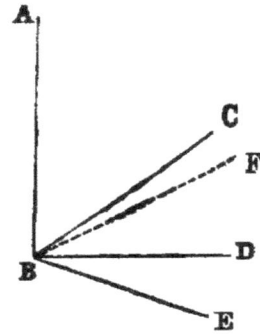

PROP. VI.—THEOREM.

If two right lines (AB, CD) be normals to the same plane (X), they shall be parallel to one another.

Dem.—Let AB, CD meet the plane X at the points B, D. Join BD, and in the plane X draw DE at right angles to BD; take any point E in DE. Join BE, AE, AD. Then because AB is normal to X, the angle ABE is right. Therefore $AE^2 = AB^2 + BE^2 = AB^2 + BD^2 + DE^2$; because the angle BDE is right. But $AB^2 + BD^2 = AD^2$, because the angle ABD is right. Hence $AE^2 = AD^2 + DE^2$. Therefore the angle ADE is right. [I. XLVIII]. And since CD is normal to the plane X, DE is perpendicular to CD. Hence DE is a common perpendicular to the three concurrent lines CD, AD, BD. Therefore these lines are coplanar [XI. V.]. But AB is coplanar with AD, BD [XI. II.]. Therefore the lines AD, BD, CD are coplanar; and since the angles ABD, BDC are right, *the line AB is parallel to CD* [I. XXVIII.].

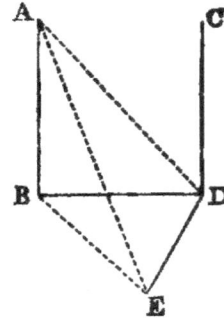

DEF. VII.—*If from every point in a given line normals be drawn to a given plane, the locus of their feet is called the projection of the given line on the plane.*

174

Exercises.

1. The projection of any line on a plane is a right line.

2. The projection on either of two intersecting planes of a normal to the other plane is perpendicular to the line of intersection of the planes.

PROP. VII.—THEOREM.

Two parallel lines (AB, CD) and any line (EF) intersecting them are coplanar.

Dem.—If possible let the intersecting line be out of the plane, as EGF. And in the plane, of the parallels draw [I. Post. II.] the right line EHF. Then we have two right lines EGF, EHF, enclosing a space, which [I. Axiom X.] is impossible. *Hence the two parallel right lines and the transversal are coplanar.*

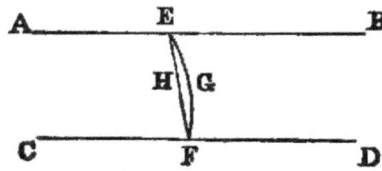

Or thus: Since the points E, F are in the plane of the parallels, the line joining these points is in that plane [I. Def. VI].

PROP. VIII.—THEOREM.

If one (AB) of two parallel right lines (AB, CD), be normal to a plane (X), the other line (CD) shall be normal to the same plane.

Dem.—Let AB, CD meet the plane X in the points B, D. Join BD. Then the lines AB, BD, CD are coplanar. Now in the plane X, to which AB is normal, draw DE at right angles to BD. Take any point E in DE, and join BE, AE, AD.

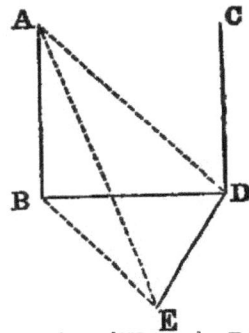

Then because AB is normal to the plane X, it is perpendicular to the line BE in that plane [XI. Def. VI.]. Hence the angle ABE is right; therefore $AE^2 = AB^2 + BE^2 = AB^2 + BD^2 + DE^2$ (because BDE is right (const.)) $= AD^2 + DE^2$ (because ABD is right (hyp.)). Therefore the angle ADE is right. Hence DE is at right angles both to AD and BD. Therefore [XI. IV.] DE is perpendicular to CD, which is coplanar and concurrent with AD and BD. Again, since AB and CD are parallel, the sum of the angles ABD, BDC is two right angles [I. XXIX.]; but ABD is right (hyp.); therefore BDC is right. Hence CD is perpendicular to the two lines DB, DE, and therefore [XI. IV.] *it is normal to their plane, that is, it is normal to X.*

175

PROP. IX—THEOREM.

Two right lines (AB, CD) which are each parallel to a third line (EF) are parallel to one another.

Dem.—If the three lines be coplanar, the Proposition is evidently the same as I. xxx. If they are not coplanar, from any point G in EF draw in the planes of EF, AB; EF, CD, respectively, the lines GH, GK each perpendicular to EF [I. xi.]. Then because EF is perpendicular to each of the lines GH, GK, it is normal to their plane [XI. iv.]. And because AB is parallel to EF (hyp.), and EF is normal to the plane GHK, AB is normal to the plane GHK [XI. viii.]. In like manner CD is normal to the plane HGK. Hence, since AB and CD are *normals to the same plane, they are parallel to one another.*

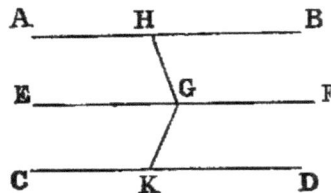

PROP. X.—THEOREM.

If two intersecting right lines (AB, BC) be respectively parallel to two other intersecting right lines (DE, EF), the angle (ABC) between the former is equal to the angle (DEF) between the latter.

Dem.—If both pairs of lines be coplanar, the proposition is the same as I. xxix., Ex. 2. If not, take any points A, C in the lines AB, BC, and cut off $ED = BA$, and $EF = BC$ [I. iii.]. Join AD, BE, CF, AC, DF. Then because AB is equal and parallel to DE, AD is equal and parallel to BE [I. xxxiii]. In like manner CF is equal and parallel to BE. Hence [XI. ix.] AD is equal and parallel to CF. Hence [I. xxxiii.] AC is equal to DF. Therefore the triangles ABC, DEF, have the three sides of one respectively equal to the three sides of the other. *Hence* [I. viii.] *the angle ABC is equal to DEF.*

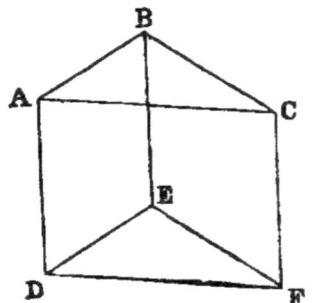

DEF. viii.—*Two planes which meet are perpendicular to each other, when the right lines drawn in one of them perpendicular to their common section are normals to the other.*

DEF. ix.—*When two planes which meet are not perpendicular to each other, their inclination is the acute angle contained by two right lines drawn from any point of their common section at right angles to it—one in one plane, and the other in the other.*

Observation.—These definitions tacitly assume the result of Props. iii. and X. of this book. On this account we have departed from the usual custom of placing them at the beginning of the book. We have altered the place of Definition vi. for a similar reason.

176

PROP. XI.—PROBLEM.

To draw a normal to a given plane (BH) from a given point (A) not in it.

Sol.—In the given plane BH draw any line BC, and from A draw AD perpendicular to BC [I. XII.]; then if AD be perpendicular to the plane, the thing required is done. If not, from D draw DE in the plane BH at right angles to BC [I. XI.], and from A draw AF [I. XII.] perpendicular to DE. *AF is normal to the plane BH.*

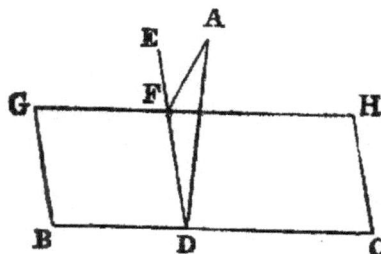

Dem.—Draw GH parallel to BC. Then because BC is perpendicular both to ED and DA, it is normal to the plane of ED, DA [XI. IV.]; and since GH is parallel to BC, it is normal to the same plane [XI. VIII.]. Hence AF is perpendicular to GH [XI. Def. VI.], and AF is perpendicular to DE (const.). *Therefore AF is normal to the plane of GH and ED—that is, to the plane BH.*

PROP. XII.—PROBLEM.

To draw a normal to a given plane from a given point (A) in the plane.

Sol.—From any point B not in the plane draw [XI. XI.] BC normal to it. If this line pass through A it is the normal required. If not, from A draw AD parallel to BC [I. XXXI.]. Then because AD and BC are parallel, and BC is normal to the plane, AD is also normal to it [XI. VIII.], and it is drawn from the given point. *Hence it is the required normal.*

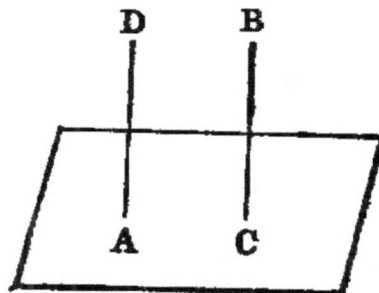

PROP. XIII.—THEOREM.

From the same point (A) there can be but one normal drawn to a given plane (X).

Dem.—1. Let A be in the given plane, and if possible let AB, AC be both normals to it, on the same side. Now let the plane of BA, AC cut the given plane X in the line DE. Then because BA is a normal, the angle BAE is right. In like manner CAE is right. Hence $BAE = CAE$, which is impossible.

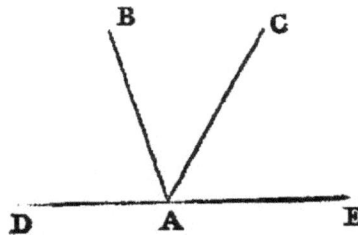

177

2. If the point be above the plane, there can be but one normal; for, if there could be two, they would be parallel [XI. VI.] to one another, which is absurd. *Therefore from the same point there can be drawn but one normal to a given plane.*

PROP. XIV.—THEOREM.

Planes (CD, EF) which have a common normal (AB) are parallel to each other.

Dem.—If the planes be not parallel, they will meet when produced. Let them meet, their common section being the line GH, in which take any point K. Join AK, BK. Then because AB is normal to the plane CD, it is perpendicular to the line AK, which it meets in that plane [XI. Def. VI.]. Therefore the angle BAK is right. In like manner the angle ABK is right. Hence the plane triangle ABK has two right angles, which is impossible. *Therefore the planes CD, EF cannot meet—that is, they are parallel.*

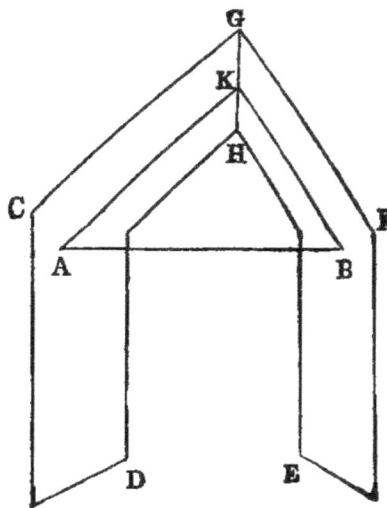

Exercises.

1. The angle between two planes is equal to the angle between two intersecting normals to these planes.

2. If a line be parallel to each of two planes, the sections which any plane passing through it makes with them are parallel.

3. If a line be parallel to each of two intersecting planes, it is parallel to their intersection.

4. If two right lines be parallel, they are parallel to the common section of any two planes passing through them.

5. If the intersections of several planes be parallel, the normals drawn to them from any point are coplanar.

PROP. XV.—THEOREM.

Two planes (AC, DF) are parallel, if two intersecting lines (AB, BC) on one of them be respectively parallel to two intersecting lines (DE, EF) on the other.

Dem.—From B draw BG perpendicular to the plane DF [XI. XI.], and let it meet that plane in G. Through G draw GH parallel to ED, and GK to EF. Now, since GH is parallel to ED (const.), and AB to ED (hyp.), AB is parallel to GH [XI. IX.]. Hence the sum of the angles ABG, BGH is two right angles [I. XXIX]; but BGH is right (const.); therefore ABG is right. In like manner

178

CBG is right. Hence *BG* is normal to the plane *AC* [XI. Def. vi.], and it is normal to *DF* (const.). Hence the planes *AC*, *DF* have a common normal *BG*; *therefore they are parallel to one another.*

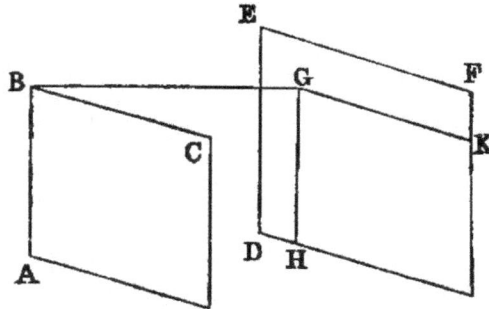

PROP. XVI.—Theorem.

If two parallel planes (AB, CD) be cut by a third plane (EF, HG), their common sections (EF, GH) with it are parallel.

Dem.—If the lines *EF*, *GH* are not parallel, they must meet at some finite distance. Let them meet in *K*. Now since *K* is a point in the line *EF*, and *EF* is in the plane *AB*, *K* is in the plane *AB*. In like manner *K* is a point in the plane *CD*. Hence the planes *AB*, *CD* meet in *K*, which is impossible, since they are parallel. *Therefore the lines EF, GH must be parallel.*

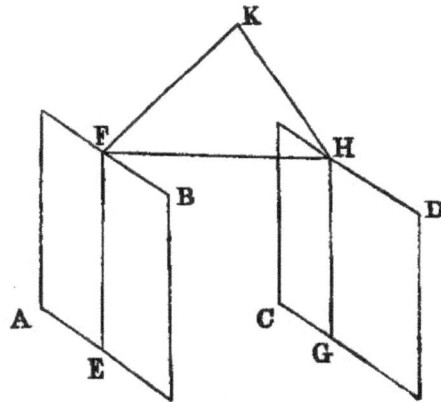

Exercises.

1. Parallel planes intercept equal segments on parallel lines.
2. Parallel lines intersecting the same plane make equal angles with it.
3. A right line intersecting parallel planes makes equal angles with them.

PROP. XVII.—Theorem.

If two parallel lines (AB, CD) be cut by three parallel planes (GH, KL, MN) in two triads of points (A, E, B; C, F, D), their segments between those points are proportional.

179

Dem.—Join AC, BD, AD. Let AD meet the plane KL in X. Join EX, XF. Then because the parallel planes KL, MN are cut by the plane ABD in the lines EX, BD, these lines are parallel [XI. XVI.]. Hence

$$AE : EB :: AX : XD \text{ [VI. II.].}$$

In like manner,

$$AX : XD :: CF : FD.$$

Therefore [V. XI.]

$$AE : EB :: CF : FD.$$

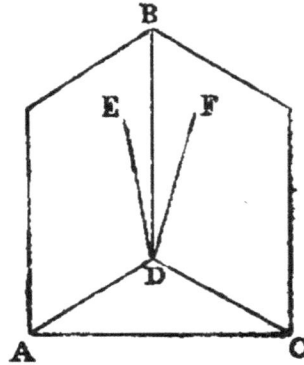

PROP. XVIII.—THEOREM.

If a right line (AB) be normal to a plane (CI), any plane (DE) passing through it shall be perpendicular to that plane.

Dem.—Let CE be the common section of the planes DE, CI. From any point F in CE draw FG in the plane DE parallel to AB [I. XXXI.]. Then because AB and FG are parallel, but AB is normal, to the plane CI; hence FG is normal to it [XI. VIII.]. Now since FG is parallel to AB, the angles ABF, BFG are equal to two right angles [I. XXIX.];

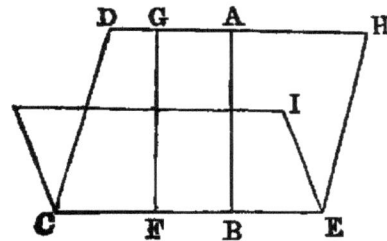

but ABF is right (hyp.); therefore BFG is right—that is, FG is perpendicular to CE. Hence every line in the plane DE, drawn perpendicular to the common section of the planes DE, CI, is normal to the plane CI. *Therefore* [XI. Def. VIII.] *the planes DE, CI are perpendicular to each other.*

PROP. XIX.—THEOREM.

If two intersecting planes (AB, BC) be each perpendicular to a third plane (ADC), their common section (BD) shall be normal to that plane.

Dem.—If not, draw from D in the plane AB the line DE perpendicular to AD, the common section of the planes AB, ADC; and in the plane BC draw BF perpendicular to the common section DC of the planes BC, ADC. Then because the plane AB is perpendicular to ADC, the line DE in AB

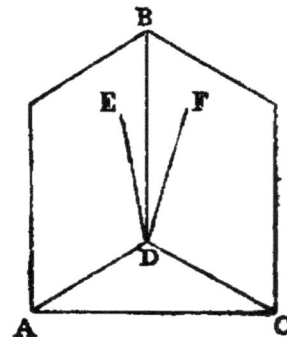

180

is normal to the plane ADC [XI. Def. VIII.]. In like manner DF is normal to it. Therefore from the point D there are two distinct normals to the plane ADC, which [XI. XIII.] is impossible. *Hence BD must be normal to the plane ADC.*

Exercises.

1. If three planes have a common line of intersection, the normals drawn to these planes from any point of that line are coplanar.

2. If two intersecting planes be respectively perpendicular to two intersecting lines, the line of intersection of the former is normal to the plane of the latter.

3. In the last case, show that the dihedral angle between the planes is equal to the rectilineal angle between the normals.

PROP. XX.—THEOREM.

The sum of any two plane angles (BAD, DAC) of a trihedral angle (A) is greater than the third (BAC).

Dem.—If the third angle BAC be less than or equal to either of the other angles the proposition is evident. If not, suppose it greater: take any point D in AD, and at the point A in the plane BAC make the angle BAE equal BAD [I. XXIII.], and cut off AE equal AD. Through E draw BC, cutting AB, AC in the points B, C. Join DB, DC.

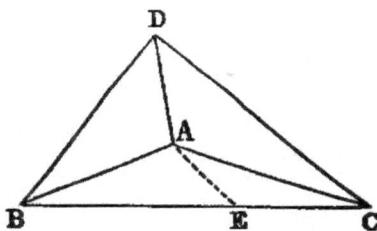

Then the triangles BAD, BAE have the two sides BA, AD in one equal respectively to the two sides BA, AE in the other, and the angle BAD equal to BAE; therefore the third side BD is equal to BE. But the sum of the sides BD, DC is greater than BC; hence DC is greater than EC. Again, because the triangles DAC, EAC have the sides DA, AC respectively equal to the sides EA, AC in the other, but the base DC greater than EC; therefore [I. XXV.] the angle DAC is greater than EAC, but the angle DAB is equal to BAE (const.). *Hence the sum of the angles BAD, DAC is greater than the angle BAC.*

PROP. XXI.—THEOREM.

The sum of all the plane angles (BAC, CAD, &c.) forming any solid angle (A) is less than four right angles.

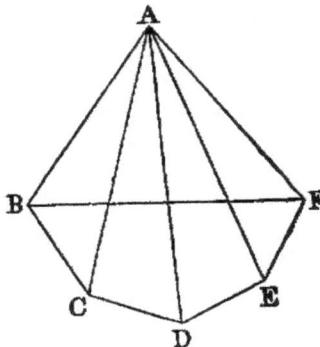

181

Dem.—Suppose for simplicity that the solid angle A is contained by five plane angles BAC, CAD, DAE, EAF, FAB; and let the planes of these angles be cut by another plane in the lines BC, CD, DE, EF, FB; then we have [XI. xx.],

$$\angle ABC + ABF \text{ greater than } FBC,$$

$$\angle ACB + ACD \qquad ,, \qquad BCD, \&\text{c.}$$

Hence, adding, we get the sum of the base angles of the five triangles BAC, CAD, &c., greater than the sum of the interior angles of the pentagon $BCDEF$ —that is, greater than six right angles. But the sum of the base angles of the same triangles, together with the sum of the plane angles BAC, CAD, &c., forming the solid angle A, is equal to twice as many right angles as there are triangles BAC, CAD, &c.—that is, equal to ten right angles. *Hence the sum of the angles forming the solid angle is less than four right angles.*

Observation.—This Prop. may not hold if the polygonal base $BCDEF$ contain *re-entrant* angles.

Exercises on Book XI.

1. Any face angle of a trihedral angle is less than the sum, but greater than the difference, of the supplements of the other two face angles.

2. A solid angle cannot be formed of equal plane angles which are equal to the angles of a regular polygon of n sides, except in the case of $n = 3$, 4, or 5.

3. Through one of two non-coplanar lines draw a plane parallel to the other.

4. Draw a common perpendicular to two non-coplanar lines, and show that it is the shortest distance between them.

5. If two of the plane angles of a tetrahedral angle be equal, the planes of these angles are equally inclined to the plane of the third angle, and conversely. If two of the planes of a trihedral angle be equally inclined to the third plane, the angles contained in those planes are equal.

6. The three lines of intersection of three planes are either parallel or concurrent.

7. If a trihedral angle O be formed by three right angles, and A, B, C be points along the edges, the orthocentre of the triangle ABC is the foot of the normal from O on the plane ABC.

8. If through the vertex O of a trihedral angle O— ABC any line OD be drawn interior to the angle, the sum of the rectilineal angles DOA, DOB, DOC is less than the sum, but greater than half the sum, of the face angles of the trihedral.

9. If on the edges of a trihedral angle O—ABC three equal lines OA, OB, OC be taken, each of these is greater than the radius of the circle described about the triangle ABC.

10. Given the three angles of a trihedral angle, find, by a plane construction, the angles between the containing planes.

11. If any plane P cut the four sides of a *Gauche* quadrilateral (a quadrilateral whose angular points are not coplanar) $ABCD$ in four points, a, b, c, d, then the product of the four ratios

$$\frac{Aa}{aB}, \quad \frac{Bb}{bC}, \quad \frac{Cc}{cD}, \quad \frac{Dd}{dA}$$

is plus unity, and conversely, if the product

$$\frac{Aa}{aB} \cdot \frac{Bb}{bC} \cdot \frac{Cc}{cD} \cdot \frac{Dd}{dA} = +1,$$

the points a, b, c, d are coplanar.

12. If in the last exercise the intersecting plane be parallel to any two sides of the quadrilateral, it cuts the two remaining sides proportionally.

DEF. X.—If at the vertex O of a trihedral angle O—ABC we draw normals OA', OB', OC' to the faces OBC, OCA, OAB, respectively, in such a manner that OA' will be on the same side of the plane OBC as OA, &c., the trihedral angle O—$A'B'C'$ is called the supplementary of the trihedral angle O—ABC.

13. If O—$A'B'C'$ be the supplementary of O—ABC, prove that O—ABC is the supplementary of O—$A'B'C'$.

14. If two trihedral angles be supplementary, each dihedral angle of one is the supplement of the corresponding face angle of the other.

15. Through a given point draw a right line which will meet two non-coplanar lines.

16. Draw a right line parallel to a given line, which will meet two non-coplanar lines.

17. Being given an angle AOB, the locus of all the points P of space, such that the sum of the projections of the line OP on OA and OB may be constant, is a plane.

183

APPENDIX.

PRISM, PYRAMID, CYLINDER, SPHERE, AND CONE

DEFINITIONS.

I. A *polyhedron* is a solid figure contained by plane figures: if it be contained by four plane figures it is called a *tetrahedron*; by six, a *hexahedron*; by eight, an *octahedron*; by twelve, a *dodecahedron*; and if by twenty, an *icosahedron*.

II. If the plane faces of a polyhedron be equal and similar rectilineal figures, it is called a *regular polyhedron.*

III. A *pyramid* is a polyhedron of which all the faces but one meet in a point. This point is called the *vertex*; and the opposite face, the *base.*

IV. A *prism* is a polyhedron having a pair of parallel faces which are equal and similar rectilineal figures, and are called its *ends*. The others, called its *side faces*, are parallelograms.

V. A prism whose ends are perpendicular to its sides is called a *right* prism; any other is called an *oblique* prism.

VI. The *altitude* of a pyramid is the length of the perpendicular drawn from its vertex to its base; and the altitude of a prism is the perpendicular distance between its ends.

VII. A *parallelopiped* is a prism whose bases are parallelograms. A parallelopiped is evidently a *hexahedron.*

VIII. A *cube* is a rectangular parallelopiped, all whose sides are squares.

IX. A *cylinder* is a solid figure formed by the revolution of a rectangle about one of its sides, which remains fixed, and which is called its *axis*. The circles which terminate a cylinder are called its *bases* or *ends.*

X. A *cone* is the solid figure described by the revolution of a right-angled triangle about one of the legs, which remains fixed, and which is called the *axis*. The other leg describes the *base*, which is a circle.

XI. A *sphere* is the solid described by the revolution of a semicircle about a diameter, which remains fixed. The *centre* of the sphere is the centre of the *generating* semicircle. Any line passing through the centre of a sphere and terminated both ways by the surface is called a *diameter*.

PROP. I.—THEOREM.

Right prisms ($ABCDE$–$FGHIJ$, $A'B'C'D'E'$–$F'G'H'I'J'$) which have bases ($ABCDE$, $A'B'C'D'E'$) that are equal and similar, and which have equal altitudes, are equal.

Dem.—Apply the bases to each other; then, since they are equal and similar figures, they will coincide— that is, the point A with A', B with B', &c. And since AF is equal to $A'F'$, and each is normal to its respective base, the: point F will coincide with F'. In the same manner the points G, H, I, J will coincide respectively with the points G', H', I', J'. Hence the prisms are equal in every respect.

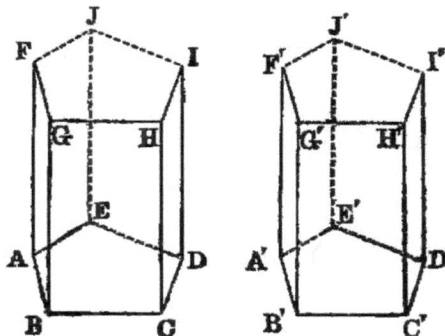

Cor. 1.—Right prisms which have equal bases (EF, $E'F'$) and equal altitudes are equal in volume.

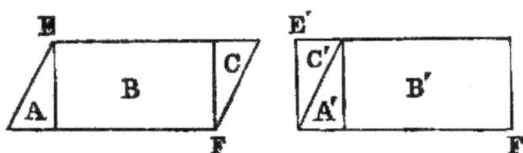

Dem.—Since the bases are equal, but not similar, we can suppose one of them, EF, divided into parts A, B, C, and re-arranged so as to make them coincide with the other [I. XXXV., note]; and since the prism on $E'F'$ can be subdivided in the same manner by planes perpendicular to the base, the proposition is evident.

Cor. 2.—The volumes of right prisms (X, Y) having equal bases are proportional to their altitudes.

For, if the altitudes be in the ratio of $m : n$, X can be divided into m prisms of equal altitudes by planes parallel to the base; then these m prisms

185

will be all equal. In like manner, Y can be divided into n equal prisms. *Hence* $X : Y :: m : n$.

Cor. 3.—In right prisms of equal altitudes the volumes are to one another as the areas of their bases. This may be proved by dividing the bases into parts so that the subdivisions will be equal, and then the volumes proportional to the number of subdivisions in their respective bases, that is, to their areas.

Cor. 4.—The volume of a rectangular parallelopiped is measured by the continued product of its three dimensions.

PROP. II.—THEOREM.

Parallelopipeds ($ABCD$–$EFGH$, $ABCD$–$MNOP$), having a common base ($ABCD$) and equal altitudes, are equal.

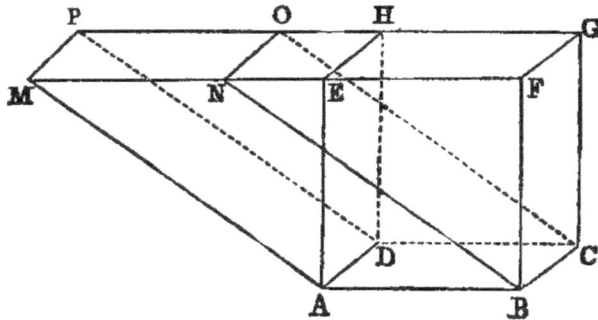

1°. Let the edges MN, EF be in one right line; then GH, OP must be in one right line. Now $EF = MN$, because each equal AB; therefore $ME = NF$; therefore the prisms AEM–DHP, and BFN–CGO, have their triangular bases AEM, BFN identically equal, and they have equal altitudes; hence they are equal; and supposing them taken away from the entire solid, *the remaining parallelopipeds $ABCD$–$EFGH$, $ABCD$–$MNOP$ are equal.*

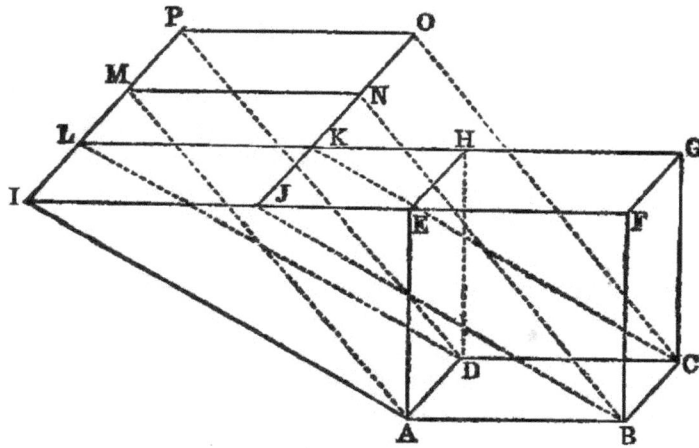

2°. Let the edges EF, MN be in different lines; then produce ON, PM to meet the lines EF and GH produced in the points J, K, L, I. Then by 1° the parallelopipeds $ABCD$–$EFGH$, $ABCD$–$MNOP$ are each equal to the parallelopiped $ABCD$–$IJKL$. Hence they an equal to one another.

Cor.—The volume of any parallelopiped is equal to the product of its base and altitude.

PROP. III.—THEOREM.

A diagonal plane of a parallelopiped divides it into two prisms of equal volume.

1°. When the parallelopiped is rectangular the proposition is evident.

2°. When it is any parallelopiped, $ABCD$–$EFGH$, the diagonal plane bisects it.

Dem.—Through the vertices A, E let planes be drawn perpendicular to the edges and cutting them in the points I, J, K; L, M, N, respectively. Then [I. XXXIV.] we have $IL = BF$, because each is equal to AE. Hence $IB = LF$. In like manner $JC = MG$. Hence the pyramid A–$IJCB$ agrees in everything but position with E–$LMGF$; hence it is equal to it in volume. To each add the solid ABC–LME, and we have the prism AIJ–ELM equal to the prism ABC–EFG. In like manner AJK–$EMN = ACD$–EGH; but (1°) the prism AIJ–$ELM = AJK$–EMN. Hence ABC–$EFG = ACD$–EGH. *Therefore the diagonal plane bisects the parallelopiped.*

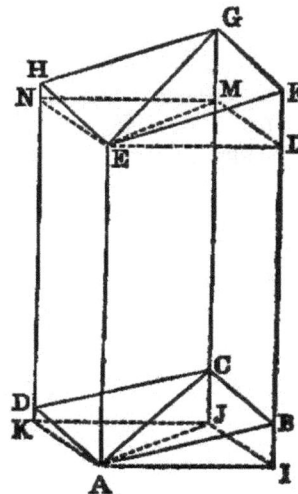

187

Cor. 1.—The volume of a triangular prism is equal to the product of its base and altitude; because it is half of a parallelopiped, which has a double base and equal altitude.

Cor. 2.—The volume of any prism is equal to the product of its base and altitude; because it can be divided into triangular prisms.

PROP. IV.—THEOREM.

If a pyramid (O–ABCDE) be cut by any plane (abcde) parallel to the base, the section is similar to the base.

Dem.—Because the plane AOB cuts the parallel planes $ABCDE$, $abcde$, the sections AB, ab are parallel [XI. XVI.] In like manner BC, bc are parallel. Hence the angle $ABC = abc$ [XI. X.]. In like manner the remaining angles of the polygon $ABCDE$ are equal to the corresponding angles of $abcde$. Again, because ab is parallel to AB, the triangles ABO, abO are equiangular.

Hence $AB : BO :: ab : bO$. [VI. IV.]

In like manner $BO : BC :: bO : bc$;

therefore $AB : BC :: ab : bc$. [*Ex æquali.*]

In like manner $BC : CD :: bc : cd$, &c.

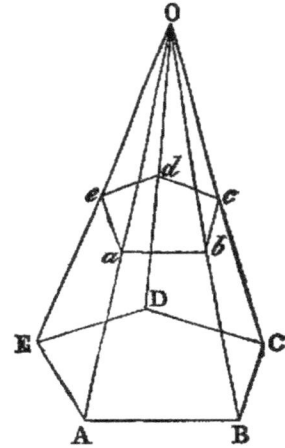

Therefore the polygons $ABCDE$, $abcde$ are equiangular, and have the sides about their equal angles proportional. *Hence they are similar.*

Cor. 1.—The edges and the altitude of the pyramid are similarly divided by the parallel plane.

Cor. 2.—The areas of parallel sections are in the duplicate ratio of the distances of their planes from the vertex.

Cor. 3.—In any two pyramids, sections parallel to their bases, which divide their altitudes in the same ratio, are proportional to their bases.

PROP. V.—THEOREM.

Pyramids (P–ABCD, p–abc), having equal altitudes (PO, po) and bases (ABCD, abc) of equal areas, have equal volumes.

188

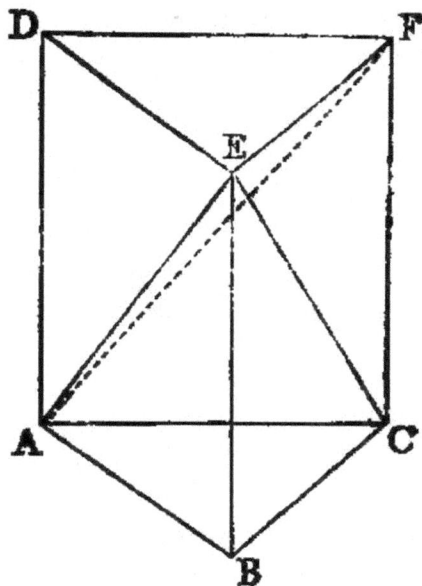

Dem.—If they be not equal in volume, let *abc* be the base of the greater; and let *ox* be the altitude of a prism, with an equal base, and whose volume is equal to their difference; then let the equal altitudes *PO*, *po* be divided into such a number of equal parts, by planes parallel to the bases of the pyramids, that each part shall be less than *ox*. Then [IV. *Cor.* 3] the sections made by these planes will be equal each to each. Now let prisms be constructed on these sections as bases and with the equal parts of the altitudes of the pyramids as altitudes, and let the prisms in *P–ABCD* be constructed below the sections, and in *p–abc*, above. Then it is evident that the sum of the prisms in *P–ABCD* is less than that pyramid, and the sum of those on the sections of *p–abc* greater than *p–abc*. Therefore the difference between the pyramids is less than the difference between the sums of the prisms, that is, less than the lower prism in the pyramid *p–abc*; but the altitude of this prism is less than *ox* (const.). Hence the difference between the pyramids is less than the prism whose base is equal to one of the equal bases, and whose altitude is equal to *ox*, and the difference is equal to this prism (hyp.), which is impossible. *Therefore the volumes of the pyramids are equal.*

Cor. 1.—*The volume of a triangular pyramid E– ABC is one third the volume of the prism ABC– DEF, having the same base and altitude.*

For, draw the plane *EAF*, then the pyramids *E– AFC*, *E–AFD* are equal, having equal bases *AFC*, *AFD*, and a common altitude; and the pyramids *E–ABC*, *F–ABC* are equal, having a common base and equal altitudes. Hence the pyramid *E–ABC* is

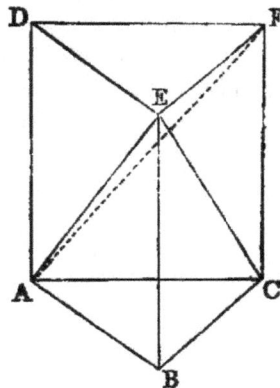

189

one of three equal pyramids into which the prism is divided. *Therefore it is one third of the prism.*

Cor. 2.—The volume of every pyramid is one-third of the volume of a prism having an equal base and altitude.

Because it may be divided into triangular pyramids by planes through the vertex and the diagonals of the base.

PROP. VI.—THEOREM.

The volume of a cylinder is equal to the product of the area of its base by its altitude

Dem.—Let O be the centre of its circular base; and take the angle AOB indefinitely small, so that the arc AB may be regarded as a right line. Then planes perpendicular to the base, and cutting it in the lines OA, OB, will be faces of a triangular prism, whose base will be the triangle AOB, and whose altitude will be the altitude of the cylinder. The volume of this prism will be equal to the area of the triangle AOB by the height of the cylinder. Hence, dividing the circle into elementary triangles, the cylinder will be equal to the sum of all the prisms, *and therefore its volume will be equal to the area of the base multiplied by the altitude.*

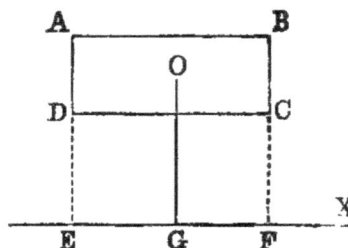

Cor. 1.—If r be the radius, and h the height of the cylinder,

$$\text{vol. of cylinder} = \pi r^2 h.$$

Cor. 2.—If $ABCD$ be a rectangle; X a line in its plane parallel to the side AB; O the middle point of the rectangle; the volume of the solid described by the revolution of $ABCD$ round X is equal to the area of $ABCD$ multiplied by the circumference of the circle described by O.

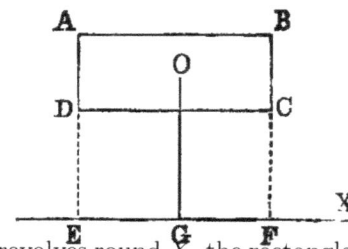

Dem.—Produce the lines AD, BC to meet X in the points E, F. Then when the rectangle revolves round X, the rectangles $ABFE$, $DCFE$ will describe cylinders whose bases will be circles having AE, DE as radii, and whose common altitude will be AB. Hence the difference between the volumes of these cylinders will be equal to the differences between the areas of the bases multiplied by AB, that is $= \pi(AE^2 - DE^2) . AB$. Therefore the volume described by $ABCD$

$$= \pi . AB . (AE + DE)(AE - DE);$$

but $\quad\quad AE + DE = 2OG$, and $AE - DE = AD.$

190

Hence volume described by the rectangle $ABCD$
$$= 2\pi \cdot OG \cdot AB \cdot AD.$$
$=$ rectangle $ABCD$ multiplied by the circumference of the circle described by its middle point O.

Observation.—This Cor. is a simple case of Guldinus's celebrated theorem. By its assistance we give in the two following corollaries original methods of finding the volumes of the cone and sphere, and it may be applied with equal facility to the solution of several other problems which are usually done by the Integral Calculus.

Cor. 3.—*The volume of a cone is one-third the volume of a cylinder having the same base and altitude.*

Dem.—Let $ABCD$ be a rectangle whose diagonal is AC. The triangle ABC will describe a cone, and the rectangle a cylinder by revolving round AB. Take two points E, F infinitely near each other in AC, and form two rectangles, EH, EK, by drawing lines parallel to AD, AB. Now if O, O' be the middle points of these rectangles, it is evident that, when the whole figure revolves round AB, the circumference of the circle described by O' will ultimately be twice the circumference of the circle described by O; and since the parallelogram EK is equal to EH [I. XLIII.], the solid described by EK (*Cor.* 1) will be equal to twice the solid described by EH. Hence, if AC be divided into an indefinite number of equal parts, and rectangles corresponding to EH, EK be inscribed in the triangles ABC, ADC, the sum of the solids described by the rectangles in the triangle ADC is equal to twice the sum of the solids described by the rectangles in the triangle ABC—that is, the difference between the cylinder and cone is equal to twice the cone. Hence the cylinder is equal to three times the cone.

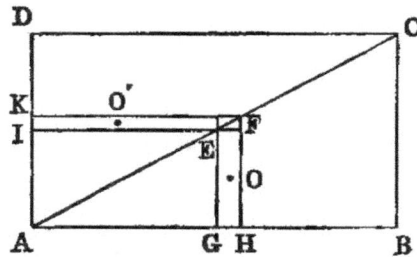

Or thus: We may regard the cone and the cylinder as limiting cases of a pyramid and prism having the same base and altitude; and since (v. *Cor.* 2) the volume of a pyramid is one-third of the volume of a prism, having the same base and altitude, the volume of the cone is one-third of the volume of the cylinder.

Cor. 4.—If r be the radius of the base of a cone, and h its height,

$$\text{vol. of cone} = \frac{\pi r^2 h}{3}.$$

Cor. 5.—*The volume of a sphere is two-thirds of the volume of a circumscribed cylinder.*

Dem.—Let AB be the diameter of the semicircle which describes the sphere; $ABCD$ the rectangle which describes the cylinder. Take two points E, F indefinitely near each other in the semicircle. Join EF, which will be a tangent, and produce it to meet the diameter PQ perpendicular to AB in N. Let R be the centre. Join RE; draw EG, FH, NL parallel to AB; and EI, FK parallel to PQ; and produce to meet LN in M and L; and let O, O' be the middle points of the rectangles EH, EK.

Now the rectangle $NG \cdot GR = PG \cdot GQ$, because each is equal to GE^2. Hence $NG : GP :: GQ : GR$, or $ME : IE :: RP + RG : RG$. Now, denoting the radii of the circles described by the points O, O' by ρ, ρ' respectively, we have ultimately $\rho = GR$ and $\rho' = \frac{1}{2}(RP + RG)$. Hence $ME : IE :: 2\rho' : \rho$; but $ME : IE ::$ rectangle EL : rectangle $EK ::$ [I. XLIII.] $EH : EK$;

$$\therefore EH : EK :: 2\rho' : \rho;$$
$$\therefore 2\pi\rho \cdot EH = 2(2\pi\rho' \cdot EK).$$

Hence the solid described by EH equal twice the solid described by EK. Therefore we infer, as in the last Cor., that the whole volume of the sphere is equal to twice the difference between the cylinder and sphere. *Therefore the sphere is two-thirds of the cylinder.*

Cor. 6.—If r be the radius of a sphere,

$$\text{vol.} = \frac{4\pi r^3}{3}.$$

192

PROP. VII.—Theorem.

The surface of a sphere is equal to the convex surface of the circumscribed cylinder.

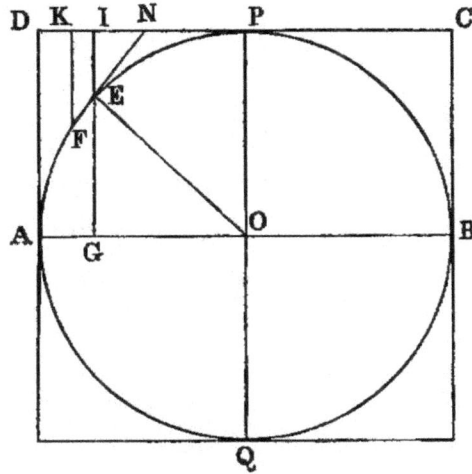

Dem.—Let AB be the diameter of the semicircle which describes the sphere. Take two points, E, F, indefinitely near each other in the semicircle. Join EF, and produce to meet the tangent CD parallel to AB in N. Draw EI, FK parallel to PQ. Produce EI to meet AB in G. Let O be the centre. Join OE.

Now we have $\qquad FE : KI :: EN : IN$ [VI. II.];

but $\qquad\qquad\qquad EN : IN :: OE : EG,$

because the triangles ENI and OEG are similar.

Hence $\qquad\qquad FE : KI :: OE : EG;$

but $\qquad\qquad\qquad OE = IG.$

Hence $EF : IK :: IG : EG$; and $IG : EG ::$ circumference of circle described by the point I : circumference of circle described by the point E. Hence the rectangle contained by EF, and circumference of circle described by E is equal to the rectangle contained by IK, and circumference of circle described by I— that is, the portion of the spherical surface described by EF is equal to the portion of the cylindrical surface described by IK. Hence it is evident, if planes be drawn perpendicular to the diameter AB—that the portions of cylindrical and spherical surface between any two of them are equal. *Hence the whole spherical surface is equal to the cylindrical surface described by CD.*

Or thus: Conceive the whole surface of the sphere divided into an indefinitely great number of equal parts, then it is evident that each of these may be regarded as the base of a pyramid having the centre of the sphere as a common vertex. Therefore the volume of the

193

sphere is equal to the whole area of the surface multiplied by one-third of the radius. Hence if S denote the surface, we have

$$S \times \frac{r}{3} = \frac{4\pi r^3}{3} \ \ [\text{VI., } Cor. \ 6];$$

therefore
$$S = 4\pi r^2.$$

That is, the area of the surface of a sphere is equal four times the area of one of its great circles.

Exercises.

1. The convex surface of a cone is equal to half the rectangle contained by the circumference of the base and the slant height.

2. The convex surface of a right cylinder is equal to the rectangle contained by the circumference of the base and the altitude.

3. If P be a point in the base ABC of a triangular pyramid O–ABC, and if parallels to the edges OA, OB, OC, through P, meet the faces in the points a, b, c, the sum of the ratios

$$\frac{Pa}{OA}, \quad \frac{Pb}{OB}, \quad \frac{Pc}{OC} = 1.$$

4. The volume of the frustum of a cone, made by a plane parallel to the base, is equal to the sum of the three cones whose bases are the two ends of the frustum, and the circle whose diameter is a mean proportional between the end diameters, and whose common altitude is equal to one-third of the altitude of the frustum.

5. If a point P be joined to the angular points A, B, C, D of a tetrahedron, and the joining lines, produced if necessary, meet the opposite faces in a, b, c, d, the sum of the ratios

$$\frac{Pa}{Aa}, \quad \frac{Pb}{Bb}, \quad \frac{Pc}{Cc}, \quad \frac{Pd}{Dd} = 1.$$

6. The surface of a sphere is equal to the rectangle by its diameter, and the circumference of a great circle.

7. The surface of a sphere is two thirds of the whole surface of its circumscribed cylinder.

8. If the four diagonals of a quadrangular prism be concurrent, it is a parallelopiped.

9. If the slant height of a right cone be equal to the diameter of its base, its total surface is to the surface of the inscribed sphere as 9 : 4.

10. The middle points of two pairs of opposite edges of a triangular pyramid are coplanar, and form a parallelogram.

11. If the four perpendiculars from the vertices on the opposite faces of a pyramid $ABCD$ be concurrent, then
$$AB^2 + CD^2 = BC^2 + AD^2 = CA^2 + BD^2.$$

12. Every section of a sphere by a plane is a circle.

13. The locus of the centres of parallel sections is a diameter of the sphere.

14. If any number of lines in space pass through a fixed point, the feet of the perpendiculars on them from another fixed point are homospheric.

15. Extend the property of Ex. 4 to the pyramid.

16. The volume of the ring described by a circle which revolves round a line in its plane is equal to the area of the circle, multiplied by the circumference of the circle described by its centre.

17. Any plane bisecting two opposite edges of a tetrahedron bisects its volume.

18. Planes which bisect the dihedral angles of a tetrahedron meet in a point.

19. Planes which bisect perpendicularly the edges of a tetrahedron meet in a point.

20. The volumes of two triangular pyramids, having a common solid angle, are proportional to the rectangles contained by the edges terminating in that angle.

21. A plane bisecting a dihedral angle of a tetrahedron divides the opposite edge into portions proportional to the areas containing that edge.

22. The volume of a sphere: the volume of the circumscribed cube as π : 6.

194

23. If h be the height, and ρ the radius of a segment of a sphere, its volume is $\dfrac{\pi h}{6}(h^2+3\rho^2)$.

24. If h be the perpendicular distance between two parallel planes, which cut a sphere in sections whose radii are ρ_1, ρ_2, the volume of the frustum is $\dfrac{\pi h}{6}\{h^2 + 3(\rho_1^2 + \rho_2^2)\}$.

25. If δ be the distance of a point P from the centre of a sphere whose radius is R, the sum of the squares of the six segments at three rectangular chords passing through P is $= 6R^2 - 2\delta^2$.

26. The volume of a sphere : the volume of an inscribed cube as $\pi : 2$.

27. If O–ABC be a tetrahedron whose angles AOB, BOC, COA are right, the square of the area of the triangle ABC is equal to the sum of the squares of the three other triangular faces.

28. In the same case, if p be the perpendicular from O on the face ABC,

$$\frac{1}{p^2} = \frac{1}{OA^2} + \frac{1}{OB^2} + \frac{1}{OC^2}.$$

29. If h be the height of an æronaut, and R the radius of the earth, the extent of surface visible $= \dfrac{2\pi R^2 h}{R+h}$.

30. If the four sides of a gauche quadrilateral touch a sphere, the points of contact are concyclic.

195

NOTES.

NOTE A.
MODERN THEORY OF PARALLEL LINES.

In every plane there is one special line called the *line at infinity*. The point where any other line in the plane cuts the line at infinity is called the *point at infinity* in that line. All other points in the line are called *finite points*. Two lines in the plane which meet the line at infinity in the same point are said to have the *same direction*, and two lines which meet it in different points to have *different directions*. Two lines which have the same direction cannot meet in any finite point [I. Axiom x.], and are parallel. Two lines which have different directions must intersect in some finite point, since, if produced, they meet the line at infinity in different points. This is a fundamental conception in Geometry, it is self-evident, and may be assumed as an Axiom (*see* Observations on the Axioms, Book I.). Hence we may infer the following general proposition:— *"Any two lines in the same plane must meet in some point in that plane; that is—(1) at infinity, when the lines have the same direction; (2) in some finite point, when they have different directions."*—See PONCELET, *Propriétés Projectives*, page 52.

NOTE B.
LEGENDRE'S AND HAMILTON'S PROOFS OF EUCLID, I. XXXII.

The discovery of the Proposition that "the sum of the three angles of a triangle is equal to two right angles" is attributed to Pythagoras. Until modern times no proof of it, independent of the theory of parallels, was known. We shall give here two demonstrations, each independent of that theory. These are due to two of the greatest mathematicians of modern times—one, the founder of the Theory of Elliptic Functions; the other, the discoverer of the Calculus of Quaternions.

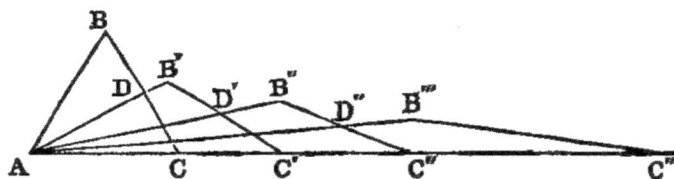

LEGENDRE'S PROOF.—Let ABC be a triangle, of which the side AC is the greatest. Bisect BC in D. Join AD. Then AD is less than AC [I. xix. Ex. 5]. Now, construct a new triangle $AB'C'$, having the side $AC' = 2AD$, and $AB' = AC$. Again, bisect $B'C'$ in D', and form another triangle $AB''C''$, having $AC'' = 2AD'$, and $AB'' = AC'$, &c. (1) The sum of the angles of the triangle ABC = the sum of the angles of $AB'C'$ [I. xvi. *Cor.* 1] = the sum of the angles of $AB''C''$ = the sum of the angles of $AB'''C'''$, &c. (2) The angle $B'AC'$ is less than half BAC; the angle $B''AC''$ is less than half $B'AC'$, and so on; hence the angle $B^{(n)}A^{(n)}$ will ultimately become infinitely small. (3) The sum of the base angles of any triangle of the series is equal to the angle of the preceding triangle (*see* Dem. I. xvi.). Hence, if the annexed diagram represent the triangle $AB^{(n+1)}C^{(n+1)}$, the sum of the base angles A and $C^{(n+1)}$ is

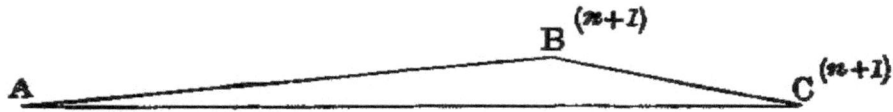

equal to the angle $B^{(n)}C^{(n)}$; and when n is indefinitely large, this angle is an infinitesimal; hence the point $B^{(n+1)}$ will ultimately be in the line AC, and the angle $AB^{(n+1)}C^{(n+1)}$ will become a straight angle [I. Def. X.], that is, it is equal to two right angles; but the sum of the angles of $AB^{(n+1)}C^{(n+1)}$ is equal to the sum of the angles ABC. *Hence the sum of the three angles of ABC is equal to two right angles.*

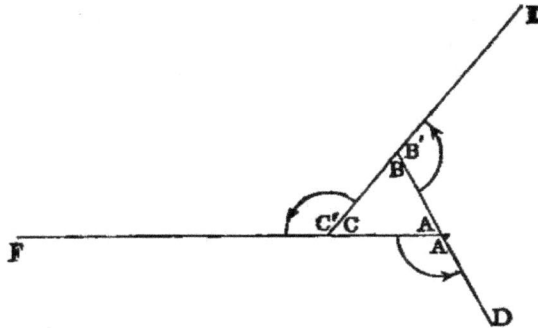

HAMILTON'S QUATERNION PROOF.—Let ABC be the triangle. Produce BA to D, and make AD equal to AC. Produce CB to E, and make BE equal to BD; finally, produce AC to F, and make CF equal to CE. Denote the exterior angles thus formed by A', B', C'. Now let the leg AC of the angle A' be turned round the point A through the angle A'; then the point C will coincide with D. Again, let the leg BD of the angle B' be turned round the point B through the angle B', until BD coincides with BE; then the point D will coincide with E. Lastly, let CE be turned round C, through the angle C', until CE coincides with CF, and the point E with F. Now, it is evident that by these rotations the point C has been brought successively into the positions D, E, F; hence, by a motion of mere translation along the line FC, the line CA can be brought into its former position. Therefore it follows, since rotation is independent of translation, that the amount of the three rotations is equal to one complete revolution round the point A; therefore $A' + B' + C' = $ four right angles; but

$$A + A' + B + B' + C + C' = \text{six right angles [I. XIII.]};$$

hence
$$A + B + C = \text{two right angles}.$$

Observation.—The foregoing demonstration is the most elementary that was ever given of this celebrated Proposition. I have reduced it to its simplest form, and without making any use of the language of Quaternions. The same method of proof will establish the more general Proposition, that the sum of the external angles of any convex rectilineal figure is equal to four right angles.

Mr. Abbott, F.T.C.D., has informed me that this demonstration was first given by Playfair in 1826, so that Hamilton was anticipated. It has been objected to on the ground that, applied verbatim to a spherical triangle, it would lead to the conclusion that the sum of the angles is two right angles, which being wrong, proves that the method is not valid. A slight consideration will show that the cases are different. In the proof given in the text there are three motions of rotation, in each of which a point describes an arc of a circle, followed by a motion of translation, in which the same point describes *a right line*, and returns to its original position. On the surface of a sphere we should have, corresponding to these, three

197

motions of rotation, in each of which the point would describe an arc of a circle, followed by a motion of rotation about the centre of the sphere, in which the point should describe *an arc of a great circle* to return to its original position. Hence, the proof for a plane triangle cannot be applied to a spherical triangle.

NOTE C.

TO INSCRIBE A REGULAR POLYGON OF SEVENTEEN SIDES IN A CIRCLE.

Analysis.—Let A be one of the angular points, AO the diameter, $A_1, A_2, \ldots A_8$ the vertices at one side of AO. Produce OA_3 to M, and OA_2 to P, making $A_3M = OA_5$, and $A_2P = OA_8$. Again, cut off $A_6N = OA_7$, and $A_1Q = OA_4$. Lastly, cut off $OR = ON$, and $OS = OQ$. Then we have [IV. Ex. 40],

$$\rho_1\rho_4 = R(\rho_3 + \rho_5) = R \cdot OM,$$
$$\rho_2\rho_8 = R(\rho_6 - \rho_7) = R \cdot ON;$$

but $$\rho_1\rho_2\rho_4\rho_8 = R^4 \qquad\qquad \text{[IV. Ex. 34]};$$

therefore $$OM \cdot ON = R^2 \qquad\qquad (1).$$

In like manner, $$OP \cdot OQ = R^2 \qquad\qquad (2).$$

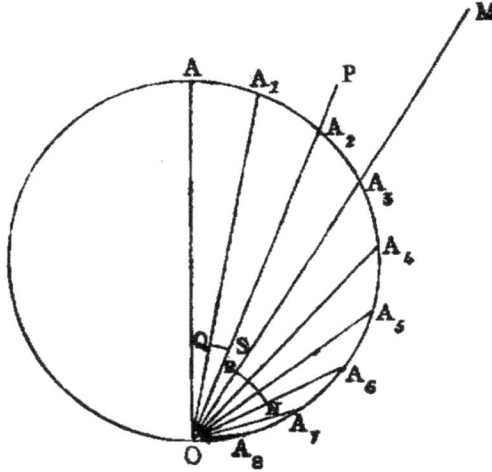

Again, $$OM \cdot ON = (\rho_3 + \rho_5)(\rho_6 - \rho_7)$$
$$= \rho_3\rho_6 + \rho_5\rho_6 - \rho_3\rho_7 - \rho_5\rho_7$$
$$= R(\rho_3 - \rho_8) + R(\rho_1 - \rho_6) - R(\rho_2 - \rho_7) - R(\rho_2 - \rho_5) \text{ [IV. Ex. 40]}.$$
$$= R(OM - ON - OP + OQ) = R(MR - PS):$$
$$MR - PS = R.$$

Again, $$MR \cdot PS = (OM - ON)(OP - OQ)$$
$$= (\rho_3 + \rho_5 - \rho_6 + \rho_7)(\rho_2 + \rho_8 - \rho_1 + \rho_4);$$

198

and performing the multiplication and substituting, we get

$$4R(OM - ON - OP + OQ) = 4R^2.$$

Hence, the rectangle and the difference of the lines MR and PS being given, each is given; hence MR is given; but $MR = OM - ON$; therefore $OM - ON$ is given; and we have proved that the rectangle $OM . ON = R^2$; therefore OM and ON are each given. In like manner, OP and OQ are each given.

Again,

$$\rho_6 . \rho_7 = R(\rho_1 - \rho_4) = R . OQ, \text{ and } \rho_6 - \rho_7 = ON.$$

Hence, since OQ and ON are each given, ρ_6 and ρ_7 are each given; therefore we can draw these chords, and we have the arc $A_6 A_7$ between their extremities given; that is, the seventeenth part of the circumference of a circle. *Hence the problem is solved.*

The foregoing analysis is due to AMPERE: see CATALAN, *Théorèmes et Problèmes de Géométrie Élémentaire*. We have abridged and simplified AMPERE's solution.

NOTE D.

TO FIND TWO MEAN PROPORTIONALS BETWEEN TWO GIVEN LINES.

The problem to find two mean proportionals is one of the most celebrated in Geometry on account of the importance which the ancients attached to it. It cannot be solved by the line and circle, but is very easy by Conic Sections. The following is a mechanical construction by the Ruler and Compass.

Sol.—Let the extremes AB, BC be placed at right angles to each other; complete the rectangle $ABCD$, and describe a circle about it. Produce DA, DC, and let a graduated ruler be made to revolve round the point B, and so adjusted that BE shall be equal to GF; then AF, CE are two mean proportionals between AB, BC.

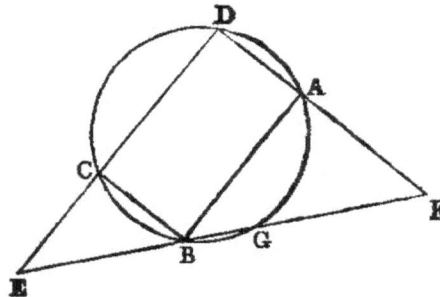

Dem.—Since BE is equal to GF, the rectangle $BE . GE = BF . GF$. Therefore $DE . CE = DF . AF$; hence $DE : DF :: AF : CE$; and by similar triangles, $AB : AF :: DE : DF$, and $CE : CB :: DE : DF$. Hence $AB : AF :: AF : CE$; and $AF : CE :: CE : CB$. Therefore AB, AF, CE, CB are continual proportions. Hence [VI. Def. IV.] AF, CE are two mean proportionals between AB and BC.

The foregoing elegant construction is due to the ancient Geometer PHILO of BYZANTIUM. If we join DG it will be perpendicular to EF. The line EF is called Philo's Line; it possesses the remarkable property of being the minimum line through the point B between the fixed lines DE, DF.

NEWTON'S CONSTRUCTION.—Let AB and L be the two given lines of which AB is the greater. Bisect AB in C. With A as centre and AC as radius, describe a circle, and in it place the chord CD equal to the second line L. Join BD, and draw by trial through A a line meeting BD, CD produced in the points E, F, so that the intercept EF will be equal to the radius of the circle. DE and FA are the mean proportionals required.

199

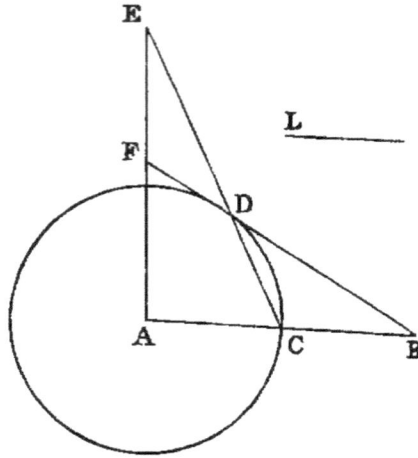

Dem.—Join AD. Since the line BF cuts the sides of the $\triangle\ ACE$, we have

$$AB.CD.EF = CB.DE.FA;\ \text{but}\ EF = CB;$$

therefore

$$AB.CD = DE.FA,\ \text{or}\ \frac{CD}{DE} = \frac{FA}{AB}.$$

Again, since the $\triangle\ ACD$ is isosceles, we have

$$ED.EC = EA^2 - AC^2 = (FA + AC)^2 - AC^2$$
$$= 2FA.AC + FA^2 = FA.AB + FA^2.$$

Hence

$$ED(ED + CD) = FA(AB + FA),$$

or

$$DE^2\left(1 + \frac{CD}{DE}\right) = FA.AB\left(1 + \frac{AF}{AB}\right),$$

therefore

$$DE^2 = FA.AB,\ \text{and we have}\ AB.CD = DE.FA.$$

Hence

$$AB,\ DE,\ FA,\ CD\ \text{are in continued proportion.}$$

NOTE E.

ON PHILO'S LINE.

I am indebted to Professor Galbraith for the following proof of the minimum property of Philo's Line. It is due to the late Professor Mac Cullagh:—*Let AC, CB be two given lines, E a fixed point, CD a perpendicular on AB; it is required to prove, if AE is equal to DB, that AB is a minimum.*

Dem.—Through E draw EM parallel to BC; make $EN = EM$; produce AB until $EP = AB$. Through the points N, P draw NT, RP each parallel to AC, and through P draw PQ parallel to BC. It is easy to see from the figure that the parallelogram QR is equal to the parallelogram MF, and is therefore given. Through P draw ST perpendicular to EP. Now, since $AE = DB$, BP is equal to DB; therefore $PS = CD$. Again, since $OP = AD$, PT is equal to CD; therefore $PS = PT$. Hence QR is the maximum parallelogram in the triangle SVT.

200

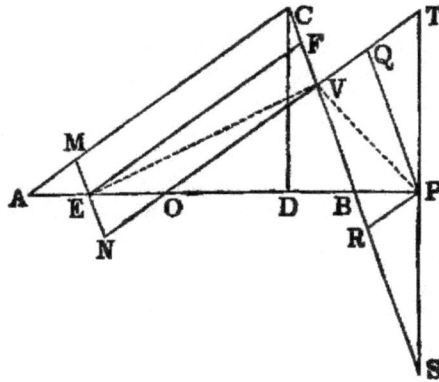

Again, if any other line $A'B$ be drawn through E, and produced to P', so that $EP' = AP'$, the point P' must fall outside ST, because the parallelogram $Q'R'$, corresponding to QR, will be equal to MF, and therefore equal to QR. Hence the line EP' is greater than EP, or $A'B'$ is greater than AB. Hence AB is a minimum.

NOTE F.

ON THE TRISECTION OF AN ANGLE.

The following mechanical method of trisecting an angle occurred to me several years ago. Apart from the interest belonging to the Problem, it is valuable to the student as a geometrical exercise:—

To trisect a given angle ACB.

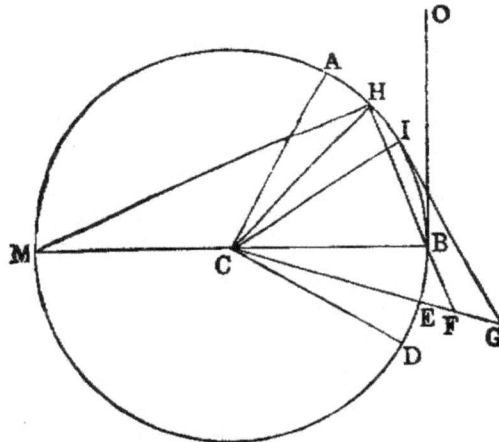

201

Sol.—Erect CD perpendicular to CA; bisect the angle BCD by CG, and make the angle ECI equal half a right angle; it is evident that CI will fall between CB and CA. Then, if we use a jointed ruler—that is two equal rulers connected by a pivot—and make CB equal to the length of one of these rulers, and, with C as centre and CB as radius, describe the circle BAM, cutting CI in I: at I draw the tangent IG, cutting CG in G.

Then, since ICG is half a right angle, and CIG is right, IGC is half a right angle; therefore IC is equal to IG; but IC equal CB; therefore $IG = CB$—equal length of one of the two equal rulers. Hence, if the rulers be opened out at right angles, and placed so that the pivot will be at I, and one extremity at C, the other at G; it is evident that the point B will be between the two rulers; then, while the extremity at C remains fixed, let the other be made to traverse the line GF, until the edge of the second ruler passes through B: it is plain that the pivot moves along the circumference of the circle. Let CH, HF, be the positions of the rulers when this happens; draw the line CH; the angle ACH is one-third of ACB.

Dem.—Produce BC to M. Join HM. Erect BO at right angles to BM. Then, because $CH = HF$, the angle $HCF = HFC$, and the angle $DCE = ECB$ (const.). Hence the angle $HCD = HBC$ [I. XXXII.], and the right angles ACD, CBO are equal; therefore the angle ACH is equal to HBO; that is [III. XXXII.], equal to HMB, or to half the angle HCB. Hence ACH is one-third of ACB.

NOTE G.

ON THE QUADRATURE OF THE CIRCLE.

Modern mathematicians denote the ratio of the circumference of a circle to its diameter by the symbol π. Hence, if r denote the radius, the circumference will be $2\pi r$; and, since the area of a circle [VI. XX. Ex. 15] is equal to half the rectangle contained by the circumference and the radius, the area will be πr^2. Hence, if the area be known, the value of π will be known; and, conversely, if the value of π be known, the area is known. On this account the determination of the value of π is called "the problem of the quadrature of the circle," and is one of the most celebrated in Mathematics. It is now known that the value of π is incommensurable; that is, that it cannot be expressed as the ratio of any two whole numbers, and therefore that it can be found only approximately; but the approximation can be carried as far as we please, just as in extracting the square root we may proceed to as many decimal places as may be required. The simplest approximate value of π was found by Archimedes, namely, 22 : 7. This value is tolerably exact, and is the one used in ordinary calculations, except where great accuracy is required. The next to this in ascending order, viz. 355 : 113, found by Vieta, is correct to six places of decimals. It differs very little from the ratio 3.1416 : 1, given in our elementary books.

Several expeditious methods, depending on the higher mathematics, are known for calculating the value of π. The following is an outline of a very simple elementary method for determining this important constant. It depends on a theorem which is at once inferred from VI., Ex. 87, namely *"If a, A denote the reciprocals of the areas of any two polygons of the same number of sides inscribed and circumscribed to a circle; a', A' the corresponding quantities for polygons of twice the number; a' is the geometric mean between a and A, and A' the arithmetic mean between a' and A."* Hence, if a and A be known, we can, by the processes of finding arithmetic and geometric means, find a' and A'. In like manner, from a', A' we can find a'', A'' related to a', A'; as a', A' are to a, A. Therefore, proceeding in this manner until we arrive at values $a^{(n)}$, $A^{(n)}$ that will agree in as many decimal places as there are in the degree of accuracy we wish to attain; and since the area of a circle is intermediate between the reciprocals of $a^{(n)}$ and $A^{(n)}$, the area of the circle can be found to any required degree of approximation.

202

If for simplicity we take the radius of the circle to be unity, and commence with the inscribed and circumscribed squares, we have

$$a = .5, \qquad A = .25.$$
$$a' = .3535533, \qquad A' = .3017766.$$
$$a'' = .3264853, \qquad A'' = .3141315.$$

These numbers are found thus: a' is the geometric mean between a and A; that is, between .5 and .25, and A' is the arithmetic mean between a' and A, or between .3535533 and .25. Again, a'' is the geometric mean between a' and A'; and A'' the arithmetic mean between a'' and A'. Proceeding in this manner, we find $a^{(13)} = .3183099$; $A^{(13)} = .3183099$. Hence the area of a circle radius unity, correct to seven decimal places, is equal to the reciprocal of .3183099; that is, equal to 3.1415926; or the value of π correct to seven places of decimals is 3.1415926. The number π is of such fundamental importance in Geometry, that mathematicians have devoted great attention to its calculation. MR. SHANKS, an English computer, carried the calculation to 707 places of decimals. The following are the first 36 figures of his result:—

$$3.141, 592, 653, 589, 793, 238, 462, 643, 383, 279, 502, 884.$$

The result is here carried far beyond all the requirements of Mathematics. Ten decimals are sufficient to give the circumference of the earth to the fraction of an inch, and thirty decimals would give the circumference of the whole visible universe to a quantity imperceptible with the most powerful microscope.

CONCLUSION.

In the foregoing Treatise we have given the Elementary Geometry of the Point, the Line, and the Circle, and figures formed by combinations of these. But it is important to the student to remark, that points and lines, instead of being distinct from, are limiting cases of, circles; and points and planes limiting cases of spheres. Thus, a circle whose radius diminishes to zero becomes a point. If, on the contrary, the circle be continually enlarged, it may have its curvature so much diminished, that any portion of its circumference may be made to differ in as small a degree as we please from a right line, and become one when the radius becomes infinite. This happens when the centre, but not the circumference, goes to infinity.

Index

THE END.

THIRD EDITION, Revised and Enlarged—3/6, cloth.

A SEQUEL

TO THE

FIRST SIX BOOKS OF THE ELEMENTS OF EUCLID.

BY

JOHN CASEY, LL.D., F.R.S.,

Fellow of the Royal University of Ireland; Vice-President, Royal Irish Academy; &c. &c.

Dublin: Hodges, Figgis, & Co.   London: Longmans, Green, & Co.

EXTRACTS FROM CRITICAL NOTICES.

"NATURE," *April* 17, 1884.

"We have noticed ('Nature,' vol. xxiv., p. 52; vol. xxvi., p. 219) two previous editions of this book, and are glad to find that our favourable opinion of it has been so convincingly indorsed by teachers and students in general. The novelty of this edition is a Supplement of Additional Propositions and Exercises. This contains an elegant mode of obtaining the circle tangential to three given circles by the methods of false positions, constructions for a quadrilateral, and a full account—for the first time in a text-book—of the Brocard, triplicate ratio, and (what the author proposes to call) the cosine circles. Dr. Casey has collected together very many properties of these circles, and, as usual with him, has added several beautiful results of his own. He has done excellent service in introducing the circles to the notice of English students.... We only need say we hope that this edition may meet with as much acceptance as its predecessors, it deserves greater acceptance."

THE "MATHEMATICAL MAGAZINE," ERIE, PENNSYLVANIA.

"Dr. Casey, an eminent Professor of the Higher Mathematics and Mathematical Physics in the Catholic University of Ireland, has just brought out a second edition of his unique 'Sequel to the First Six Books of Euclid,' in which he has contrived to arrange and to pack more geometrical gems than have appeared in any single text-book since the days of the self-taught Thomas Simpson. 'The principles of Modern Geometry contained in the work are, in the present state of Science, indispensable in Pure and Applied Mathematics, and in Mathematical Physics; and it is important that the student should become early acquainted with them.'

"Eleven of the sixteen sections into which the work is divided exhibit most excellent specimens of geometrical reasoning and research. These will be found to furnish very neat models for systematic methods of study. The other five sections contain 261 choice problems for solution. Here the earnest student will find all that he needs to bring himself abreast

209

with the amazing developments that are being made almost daily in the vast regions of Pure and Applied Geometry. On pp. 152 and 153 there is an elegant solution of the celebrated Malfatti's Problem.

"As our space is limited, we earnestly advise every lover of the 'Bright Seraphic Truth' and every friend of the 'Mathematical Magazine' to procure this invaluable book without delay."

THE "SCHOOLMASTER."

"This book contains a large number of elementary geometrical propositions not given in Euclid, which are required by every student of Mathematics. Here are such propositions as that *the three bisectors of the sides of a triangle are concurrent*, needed in determining the position of the centre of gravity of a triangle; propositions in the circle needed in Practical Geometry and Mechanics; properties of the centres of similitudes, and the theories of inversion and reciprocations so useful in certain electrical questions. The proofs are always neat, and in many cases exceedingly elegant."

THE "EDUCATIONAL TIMES."

"We have certainly seen nowhere so good an introduction to Modern Geometry, or so copious a collection of those elementary propositions not given by Euclid, but which are absolutely indispensable for every student who intends to proceed to the study of the Higher Mathematics. The style and general get up of the book are, in every way, worthy of the 'Dublin University Press Series,' to which it belongs."

THE "SCHOOL GUARDIAN."

"This book is a well-devised and useful work. It consists of propositions supplementary to those of the first six books of Euclid, and a series of carefully arranged exercises which follow each section. More than half the book is devoted to the Sixth Book of Euclid, the chapters on the 'Theory of Inversion' and on the 'Poles and Polars' being especially good. Its method skilfully combines the methods of old and modern Geometry; and a student well acquainted with its subject-matter would be fairly equipped with the geometrical knowledge he would require for the study of any branch of physical science."

THE "PRACTICAL TEACHER."

"Professor Casey's aim has been to collect within reasonable compass all those propositions of Modern Geometry to which reference is often made, but which are as yet embodied nowhere.... We can unreservedly give the highest praise to the matter of the book. In most cases the proofs are extraordinarily neat.... The notes to the Sixth Book are the most satisfactory. Feuerbach's Theorem (the nine-points circle touches inscribed and escribed circles) is favoured with two or three proofs, all of which are elegant. Dr. Hart's extension of it is extremely well proved.... We shall have given sufficient commendation to the book when we say, that the proofs of these (Malfatti's Problem, and Miquel's Theorem), and equally complex problems, which we used to shudder to attack, even by the powerful weapons of analysis, are easily and triumphantly accomplished by Pure Geometry.

"After showing what great results this book has accomplished in the minimum of space, it is almost superfluous to say more. Our author is almost alone in the field, and for the present need scarcely fear rivals."

The "Academy."

"Dr. Casey is an accomplished geometer, and this little book is worthy of his reputation. It is well adapted for use in the higher forms of our schools. It is a good introduction to the larger works of Chasles, Salmon, and Townsend. It contains both a text and numerous examples."

The "Journal of Education."

"Dr. Casey's 'Sequel to Euclid' will be found a most valuable work to any student who has thoroughly mastered Euclid, and imbibed a real taste for geometrical reasoning.... The higher methods of pure geometrical demonstration, which form by far the larger and more important portion, are admirable; the propositions are for the most part extremely well given, and will amply repay a careful perusal to advanced students."

PREFACE.

Frequent applications having been made to DR. CASEY requesting him to publish a "Key" containing the Solutions of the Exercises in his "Elements of Euclid," but his professorial and other duties scarcely leaving him any time to devote to it, I undertook, under his direction, the task of preparing one. Every Solution was examined and approved of by him before writing it for publication, so that the work may be regarded as virtually his.

The Exercises are a joint selection made by him and the late lamented Professor Townsend, S.F.T.C.D., and form one of the finest collections ever published.

JOSEPH B. CASEY.

86, SOUTH CIRCULAR-ROAD,
    December 23, 1886.

212

Price 4/6, post free.]

THE FIRST SIX BOOKS

OF THE

ELEMENTS OF EUCLID,

With Copious Annotations and Numerous Exercises.

BY

JOHN CASEY, LL.D., F.R.S.,

Fellow of the Royal University of Ireland; Vice-President, Royal Irish Academy; &c.
&c.

Dublin: Hodges, Figgis, & Co.   London: Longmans, Green, & Co.

OPINIONS OF THE WORK.

The following are a few of the Opinions received by Dr. Casey on this Work:—

"Teachers no longer need be at a loss when asked which of the numerous 'Euclids' they recommend to learners. Dr. Casey's will, we presume, supersede all others."—THE DUBLIN EVENING MAIL.

"Dr. Casey's work is one of the best and most complete treatises on Elementary Geometry we have seen. The annotations on the several propositions are specially valuable to students."—THE NORTHERN WHIG.

"His long and successful experience as a teacher has eminently qualified Dr. Casey for the task which he has undertaken.... We can unhesitatingly say that this is the best edition of Euclid that has been yet offered to the public."—THE FREEMAN'S JOURNAL.

From the REV. R. TOWNSEND, F.T.C.D., &c.

"I have no doubt whatever of the general adoption of your work through all the schools of Ireland immediately, and of England also before very long."

From GEORGE FRANCIS FITZGERALD, Esq., F.T.C.D.

"Your work on Euclid seems admirable, and is a great improvement in most ways on its predecessors. It is a great thing to call the attention of students to the innumerable variations in statement and simple deductions from propositions.... I should have preferred some modification of Euclid to a reproduction, but I suppose people cannot be got to agree to any."

213

From H. J. Cooke, Esq., The Academy, Banbridge.

"In the clearness, neatness, and variety of demonstrations, it is far superior to any text-book yet published, whilst the exercises are all that could be desired."

From James A. Poole, M.A., 29, Harcourt-street, Dublin.

"This work proves that Irish Scholars can produce Class-books which even the Head Masters of English Schools will feel it a duty to introduce into their establishments."

From Professor Leebody, Magee College, Londonderry.

"So far as I have had time to examine it, it seems to me a very valuable addition to our text-books of Elementary Geometry, and a most suitable introduction to the 'Sequel to Euclid,' which I have found an admirable book for class teaching."

From Mrs. Bryant, F.C.P., Principal of the North London Collegiate School for Girls.

"I am heartily glad to welcome this work as a substitute for the much less elegant text-books in vogue here. I have begun to use it already with some of my classes, and find that the arrangement of exercises after each proposition works admirably."

From the Rev. J. E. Reffé, French College, Blackrock.

"I am sure you will soon be obliged to prepare a Second Edition. I have ordered fifty copies more of the Euclid (this makes 250 copies for the French College). They all like the book here."

From the Nottingham Guardian.

"The edition of the First Six Books of Euclid by Dr. John Casey is a particularly useful and able work.... The illustrative exercises and problems are exceedingly numerous, and have been selected with great care. Dr. Casey has done an undoubted service to teachers in preparing an edition of Euclid adapted to the development of the Geometry of the present day."

From the Leeds Mercury.

"There is a simplicity and neatness of style in the solution of the problems which will be of great assistance to the students in mastering them.... At the end of each proposition there is an examination paper upon it, with deductions and other propositions, by means of which the student is at once enabled to test himself whether he has fully grasped the principles involved.... Dr. Casey brings at once the student face to face with the difficulties to be encountered, and trains him, stage by stage, to solve them."

From the Practical Teacher.

"The preface states that this book 'is intended to supply a want much felt by Teachers at the present day–the production of a work which, while giving the unrivalled original in all its integrity, would also contain the modern conceptions and developments of the portion of Geometry over which the elements extend.'

"The book is all, and more than all, it professes to be.... The propositions suggested are such as will be found to have most important applications, and the methods of proof are both simple and elegant. We know no book which, within so moderate a compass, puts the student in possession of such valuable results.

"The exercises left for solution are such as will repay patient study, and those whose solution are given in the book itself will suggest the methods by which the others are to be demonstrated. We recommend everyone who wants good exercises in Geometry to get the book, and study it for themselves."

214